AF356130

Remo Ruffini Festschrift

Remo Ruffini Festschrift

Editors

Remo Ruffini
Jorge Armando Rueda Hernández
Narek Sahakyan
Gregory Vereshchagin

Basel • Beijing • Wuhan • Barcelona • Belgrade • Novi Sad • Cluj • Manchester

Editors
Remo Ruffini
ICRANet Headquarters
ICRANet
Pescara
Italy

Jorge Armando Rueda Hernández
ICRANet-Ferrara
Universita' degli Studi
di Ferrara
Ferrara
Italy

Narek Sahakyan
ICRANet-Armenia
ICRANet
Yerevan
Armenia

Gregory Vereshchagin
ICRANet Headquarters
ICRANet
Pescara
Italy

Editorial Office
MDPI
St. Alban-Anlage 66
4052 Basel, Switzerland

This is a reprint of articles from the Special Issue published online in the open access journal *Universe* (ISSN 2218-1997) (available at: https://www.mdpi.com/journal/universe/special_issues/ J0M337731D).

For citation purposes, cite each article independently as indicated on the article page online and as indicated below:

Lastname, A.A.; Lastname, B.B. Article Title. *Journal Name* **Year**, *Volume Number*, Page Range.

ISBN 978-3-7258-0564-8 (Hbk)
ISBN 978-3-7258-0563-1 (PDF)
doi.org/10.3390/books978-3-7258-0563-1

Contents

Preface

Remo Ruffini received his doctorate degree from Sapienza University of Rome in 1967. He has taught in Hamburg, Princeton University, the Institute for Advanced Study, Japan, China (at USTC), Australia, and CBPF (Brazil). He is a co-author of more than 650 scientific publications and 13 books. One of his significant studies concerned boson stars, named "Introducing the Black Hole" with J.A. Wheeler, which focused on the limiting critical mass of NS. He identified the first black hole (BH) in our galaxy (Cignus X-1) using UHURU satellite data with Riccardo Giacconi, and consequently, he received the Cressy Morrison Award in 1973. He returned to Sapienza University in 1978 and promoted a Rome–Stanford collaboration on gravitational-wave detectors. Together with the European, US, and Chinese institutions, he established ICRA in 1985 and ICRANet in Italy, Armenia, France, and Brazil in 2005. He developed the understanding of GRBs, confirmed by the largest telescopes on Earth, from their discovery in 1973 to their cosmological origin in 1997 and to the determination of seven different GRB families and their conceptual understanding in 2018. ICRA and ICRANet have always been characterized by their ability to establish and maintain international relations in the scientific field, as can be deduced from their publications.

A conference celebrating Remo Ruffini's 80th birthday was held in Nice, France, from 16 to 18 May 2022, with the participation of over 90 scientists. Among the contributions presented orally, there were those by Rashid Sunyaev, Peter Predehl, Demetrios Christodoulou, Thibault Damour, Nathalie Deruelle, Roy Kerr, Tsvi Piran, Claus Laemmerzahl, Asghar Qadir, Chen Pisin, and Marco Tavani, in the presence of Agnès Rampal, the representative of the Mayor of Nice. An extraordinary moment of the meeting in Nice was the delivery of the Marcel Grossmann Award to Rashid Sunyaev and Peter Predehl for the Spectr-Roentgen-Gamma (SRG) mission.

This volume contains 11 papers by some of the meeting participants and collaborators of Remo Ruffini.

Remo Ruffini, Jorge Armando Rueda Hernández, Narek Sahakyan, and Gregory Vereshchagin

Editors

 *universe*  MDPI

Article

Spherically Symmetric C^3 Matching in General Relativity

Hernando Quevedo

Instituto de Ciencias Nucleares, Universidad Nacional Autónoma de México, AP 70543, Ciudad de México 04510, Mexico; quevedo@nucleares.unam.mx

Abstract: We study the problem of matching interior and exterior solutions to Einstein's equations along a particular hypersurface. We present the main aspects of the C^3 matching approach that involve third-order derivatives of the corresponding metric tensors in contrast to the standard C^2 matching procedures known in general relativity, which impose conditions on the second-order derivatives only. The C^3 alternative approach does not depend on coordinates and allows us to determine the matching surface by using the invariant properties of the eigenvalues of the Riemann curvature tensor. As a particular example, we apply the C^3 procedure to match the exterior Schwarzschild metric with a general spherically symmetric interior spacetime with a perfect fluid source and obtain that on the matching hypersurface, the density and pressure should vanish, which is in accordance with the intuitive physical expectation.

Keywords: exact solutions; matching conditions; curvature eigenvalues

1. Introduction

One important problem in astrophysics consists of describing the gravitational field of compact objects. Consider the gravitational field of a compact object, whose surface is denoted as Σ. Let U^+ and U^- represent the Newtonian gravitational potential outside and inside the object, respectively. This means that the potentials should be solutions to the Poisson equation (in Cartesian coordinates)

$$\Delta U \;= \left(\frac{\partial^2}{\partial x^2} + \frac{\partial^2}{\partial y^2} + \frac{\partial^2}{\partial z^2} \right) U^- = 4\pi G \rho, \tag{1}$$

and the Laplace equation

$$\Delta U^+ = 0, \tag{2}$$

respectively. Here, ρ is the density of the matter distribution that generates the gravitational field. The exterior (interior) potential describes the field outside (inside) the mass distribution. In general, the problem of finding solutions to the Laplace and Poisson equations is considered in the framework of potential theory. Usually, it is assumed that the mass distribution fulfills certain symmetry conditions that allow us to simplify the complexity of the corresponding differential equations. Consider, for instance, the case of a spherically symmetric mass distribution with radius R. Then, the Laplace equation reduces to

$$\Delta U^+ = \frac{1}{r^2} \frac{\partial}{\partial r} \left(r^2 \frac{\partial U^+}{\partial r} \right) = 0 \,, \tag{3}$$

where r is the radial coordinate, and the solution for the exterior potential can be expressed as

$$U^+ = \frac{M}{r} \,, \tag{4}$$

Citation: Quevedo, H. Spherically Symmetric C^3 Matching in General Relativity. *Universe* **2023**, *9*, 419. https://doi.org/10.3390/ universe9090419

Academic Editor: Guillermo A. Mena Marugán

Received: 22 June 2023
Revised: 12 September 2023
Accepted: 12 September 2023
Published: 14 September 2023

where M is a constant of integration. The corresponding Poisson equation for an arbitrary function $\rho(r)$ can be solved by using the Green function

$$U^- = -G \int \frac{\rho(r')}{|r - r'|} dr' . \tag{5}$$

The matching consists in demanding that on the surface Σ, which in this case corresponds to $r = R$, the potentials U^- and U^+ coincide. This is easily done, and we obtain as a result that M can be written in terms of $\rho(r)$ and corresponds to the total mass of the body.

Another practical example is that of an axially symmetric mass distribution. In this case, the Laplace equation becomes

$$\Delta U^- = \frac{1}{r^2} \frac{\partial}{\partial r} \left(r^2 \frac{\partial U^-}{\partial r} \right) + \frac{1}{r^2 \sin \theta} \frac{\partial}{\partial \theta} \left(\sin \theta \frac{\partial U^-}{\partial \theta} \right) = 0 , \tag{6}$$

where θ is the azimuthal angle. This is a linear differential equation whose general solution can be represented as

$$U^+ = \sum_{n=0}^{\infty} \frac{a_n}{r^{\frac{n+1}{2}}} P_n(\cos \theta) , \tag{7}$$

where $P_n(\cos \theta)$ are the Legendre polynomials and a_n are constants. As for the internal potential U^-, using the Green function formalism, the solution can be expressed as an infinite series, each term of which represents a particular multipole moment. The matching consists in demanding that the interior and exterior potentials coincide on Σ. This can be reached by calculating the explicit value of the series of the interior potential, which is given in terms of the density of the mass distribution, and demanding that it coincides term by term with the exterior potential (7) on Σ. As a result, the exterior multipoles become fixed by the values of the interior multipoles on the matching surface. This means that in Newtonian gravity, the matching problem can be solved uniquely by using multipole moments.

Consider now the matching problem in Einstein's theory of gravity, where the gravitational field of a mass distribution must be described by a metric $g_{\mu\nu}$ ($\mu, \nu = 0, 1, 2, 3$), satisfying Einstein's equations,

$$R_{\mu\nu} - \frac{1}{2} g_{\mu\nu} R = 8\pi T_{\mu\nu} , \tag{8}$$

in the interior part of the mass ($T_{\mu\nu} \neq 0$) as well as outside in empty space ($T_{\mu\nu} = 0$). For concreteness, let us consider a mass distribution whose internal structure is described by a perfect fluid

$$T_{\mu\nu} = (\rho + p) u_\mu u_\nu + p g_{\mu\nu} , \tag{9}$$

where $\rho(x^\mu)$ is the density, $p(x^\mu)$ is the pressure, and u^μ is the 4-velocity of a particle inside the fluid. As in Newtonian gravity, the complexity of the corresponding partial differential equations can be reduced by imposing symmetry conditions on the gravitational source. For instance, if we limit ourselves to spherically symmetric gravitational fields, the field equations reduce to a set of ordinary differential equations that can be solved analytically. Furthermore, in the case of vacuum gravitational fields, the differential equations can be solved in general and, by virtue of Birkhoff's theorem [1], the solution turns out to be unique and is known as the Schwarzschild spacetime [2], which describes the gravitational field of a static, spherically symmetric mass distribution. In the case of the interior gravitational field of compact objects, the situation is much more complicated. In the literature, there exists a reasonable number of interior spherically symmetric solutions [3], which are candidates to be matched with the exterior Schwarzschild metric. In this work, we will obtain the conditions under which the Schwarzschild spacetime can be matched with an interior spherically symmetric perfect-fluid solution. To this end, we will apply the C^3 approach, which is based upon the use of the eigenvalues of the curvature tensor.

The matching problem in general relativity has been the subject of intensive research since Darmois published in 1927 his matching method [4], stating that the first and second fundamental forms should be continuous across the matching surface and implying conditions on the second derivatives of the metric. For this reason, this method is usually called C^2 matching. However, as pointed out by Israel in [5], in practice, the C^2 matching is of limited utility because it requires the use of particular sets of admissible coordinates. Israel also proposed a generalization of Darmois conditions to include the more realistic case in which surface discontinuities are present. This generalization yields the thin-shell approach, which is widely used in the literature.

The C^3 alternative approach is different. We demand that the eigenvalues of the curvature tensor be continuous across the matching surface and use their derivatives to determine the location where the matching can be performed. Since the curvature eigenvalues are scalars, the C^3 matching conditions are invariant. In addition, it is also possible to generalize the C^3 method to include the case of surface discontinuities by using Israel's thin-shell proposal.

This work is organized as follows. In Section 2, we present a method to compute the eigenvalues of the Riemann curvature tensor, which is based upon the use of a local orthonormal basis and the formalism of differential forms with Cartan's equations as the underlying structure. Then, in Section 3, we describe the C^3 method and present the corresponding matching conditions. In Section 4, we apply the C^3 matching procedure in the case of spherically symmetric spacetimes. Finally, in Section 5, we discuss our results and comment on further applications of the C^3 matching.

2. Eigenvalues of the Riemann Curvature Tensor

The eigenvalues of the curvature tensor can be computed in different ways [6]. Here, we use the formalism of differential forms with a set of local orthonormal tetrads. From a physical point of view, an observer would choose a local orthonormal tetrad as the simplest and most natural frame of reference. Indeed, according to the equivalence principle, local measurements of space and time can be performed in a gravity-free environment so it is natural to use locally the flat Minkowski metric of special relativity. On the other hand, the use of local tetrads allows us to perform measurements that are invariant with respect to coordinate transformations. The only freedom remaining in the choice of this local frame is a Lorentz transformation. So, let us choose the orthonormal tetrad as

$$ds^2 = g_{\mu\nu}dx^\mu \otimes dx^\nu = \eta_{ab}\vartheta^a \otimes \vartheta^b \,, \tag{10}$$

with $\eta_{ab} = \mathrm{diag}(-1,1,1,1)$, and $\vartheta^a = e^a_\mu dx^\mu$. Then, the tetrad components ϑ^a can be interpreted as differential one-forms. Furthermore, Cartan's first structure equation

$$d\vartheta^a = -\omega^a{}_b \wedge \vartheta^b \tag{11}$$

can be used to determine explicitly the components of the connection one-form $\omega^a{}_b$, which, in turn, are used to define the curvature two-form $\Omega^a{}_b$ by means of Cartan's second structure equation

$$\Omega^a{}_b = d\omega^a{}_b + \omega^a{}_c \wedge \omega^c{}_b = \frac{1}{2}R^a{}_{bcd}\vartheta^c \wedge \vartheta^d \,, \tag{12}$$

where $R^a{}_{bcd}$ are the components of the Riemann curvature tensor in the local orthonormal frame ϑ^a.

The curvature tensor can be represented as a (6×6)-matrix by introducing the bivector indices $A,..., A = 1,\ldots 6$, which encode the information of two different tetrad indices, i.e., $ab \to A$. A particular choice of this correspondence is [1]

$$01 \to 1\,, \quad 02 \to 2\,, \quad 03 \to 3\,, \quad 23 \to 4\,, \quad 31 \to 5\,, \quad 12 \to 6\,. \tag{13}$$

Then, the Riemann tensor can be represented by the symmetric matrix $\mathbf{R}_{AB}$ with

$$\mathbf{R}_{AB} = \begin{pmatrix} R_{0101} & R_{0102} & R_{0103} & R_{0123} & R_{0131} & R_{0112} \\ R_{0102} & R_{0202} & R_{0203} & R_{0223} & R_{0231} & R_{0212} \\ R_{0103} & R_{0203} & R_{0303} & R_{0323} & R_{0331} & R_{0312} \\ R_{0123} & R_{0223} & R_{0323} & R_{2323} & R_{2331} & R_{1223} \\ R_{0131} & R_{0231} & R_{0331} & R_{2331} & R_{3131} & R_{1231} \\ R_{0112} & R_{0212} & R_{0312} & R_{1223} & R_{1231} & R_{1212} \end{pmatrix}, \tag{14}$$

which possesses 21 independent components. However, the first Bianchi identity

$$R_{a[bcd]} = 0 \Leftrightarrow R_{0123} + R_{0312} + R_{0231} = 0, \tag{15}$$

which in bivector representation reads

$$\mathbf{R}_{14} + \mathbf{R}_{25} + \mathbf{R}_{36} = 0, \tag{16}$$

imposes an additional relationship between the components of the curvature matrix and, consequently, reduces the number of independent components to 20, as it should be in the case of a 4-dimensional Riemannian manifold.

We now consider Einstein's equations with cosmological constant in the orthonormal frame ϑ^a,

$$R_{ab} - \frac{1}{2}R\eta_{ab} + \Lambda\eta_{ab} = \kappa T_{ab}, \quad R_{ab} = R^c{}_{acb}, \tag{17}$$

which represent a relationship between the components of the curvature tensor, the cosmological constant, and the components of the energy-momentum tensor.

By writing the Ricci tensor R_{ab} and the curvature scalar R explicitly in terms of the components of the Riemann tensor in the bivector representation, Einstein's equations reduce to a set of ten algebraic equations that relate the components of the matrix $\mathbf{R}_{AB}$. This means that we can express ten of the components $\mathbf{R}_{AB}$ in terms of the remaining ten components. For concreteness, we choose as independent components the following: $\mathbf{R}_{11}$, $\mathbf{R}_{12}$, $\mathbf{R}_{13}$, $\mathbf{R}_{14}$, $\mathbf{R}_{15}$, $\mathbf{R}_{16}$, $\mathbf{R}_{22}$, $\mathbf{R}_{23}$, $\mathbf{R}_{25}$, and $\mathbf{R}_{26}$. Introducing the resulting equations into the matrix $\mathbf{R}_{AB}$, only ten components remain independent. Then, the curvature matrix can be represented as [7,8]

$$\mathbf{R}_{AB} = \begin{pmatrix} \mathbf{M}_1 & \mathbf{L} \\ \mathbf{L} & \mathbf{M}_2 \end{pmatrix}, \tag{18}$$

where

$$\mathbf{L} = \begin{pmatrix} \mathbf{R}_{14} & \mathbf{R}_{15} & \mathbf{R}_{16} \\ \mathbf{R}_{15} - \kappa T_{03} & \mathbf{R}_{25} & \mathbf{R}_{26} \\ \mathbf{R}_{16} + \kappa T_{02} & \mathbf{R}_{26} - \kappa T_{01} & -\mathbf{R}_{14} - \mathbf{R}_{25} \end{pmatrix},$$

and $\mathbf{M}_1$ and $\mathbf{M}_2$ are 3×3 symmetric matrices

$$\mathbf{M}_1 = \begin{pmatrix} \mathbf{R}_{11} & \mathbf{R}_{12} & \mathbf{R}_{13} \\ \mathbf{R}_{12} & \mathbf{R}_{22} & \mathbf{R}_{23} \\ \mathbf{R}_{13} & \mathbf{R}_{23} & -\mathbf{R}_{11} - \mathbf{R}_{22} - \Lambda + \kappa\left(\frac{T}{2} + T_{00}\right) \end{pmatrix},$$

$$\mathbf{M}_2 = \begin{pmatrix} -\mathbf{R}_{11} + \kappa\left(\frac{T}{2} + T_{00} - T_{11}\right) & -\mathbf{R}_{12} - \kappa T_{12} & -\mathbf{R}_{13} - \kappa T_{13} \\ -\mathbf{R}_{12} - \kappa T_{12} & -\mathbf{R}_{22} + \kappa\left(\frac{T}{2} + T_{00} - T_{22}\right) & -\mathbf{R}_{23} - \kappa T_{23} \\ -\mathbf{R}_{13} - \kappa T_{13} & -\mathbf{R}_{23} - \kappa T_{23} & \mathbf{R}_{11} + \mathbf{R}_{22} + \Lambda - \kappa T_{33} \end{pmatrix},$$

where T is the trace of the energy-momentum tensor, $T = \eta^{ab}T_{ab}$. Accordingly, this is the most general form of a curvature tensor that satisfies Einstein's equations with cosmological constant and arbitrary energy-momentum tensor.

We note that the traces of the above matrices turn out to be of particular importance. Indeed,

$$\mathrm{Tr}(\mathbf{L}) = 0\,, \tag{19}$$

$$\mathrm{Tr}(\mathbf{M_1}) = -\Lambda + \kappa\left(\frac{T}{2} + T_{00}\right), \quad \mathrm{Tr}(\mathbf{M_2}) = +\Lambda + \kappa T_{00}\,. \tag{20}$$

As shown above, the first equation follows from the Bianchi identities. The second and third equations can be proved by direct computation. Consequently, the trace of the curvature matrix can be expressed as

$$\mathrm{Tr}(\mathbf{R_{AB}}) = \kappa\left(\frac{T}{2} + 2T_{00}\right). \tag{21}$$

Thus, we see that all the relevant traces depend on the components of the energy-momentum tensor only.

The eigenvalues of the curvature tensor correspond to the eigenvalues of the matrix $\mathbf{R}_{AB}$. In general, they are functions λ_i, with $i = 1, 2, ..., 6$, which depend on the parameters and coordinates entering the tetrads ϑ^a.

As a particular example of the bivector representation of the curvature, consider now the case of a perfect fluid energy-momentum tensor with density ρ and pressure p, i.e.,

$$T_{ab} = (\rho + p)u_a u_b + p\eta_{ab}\,, \tag{22}$$

where $u^a = (-1, 0, 0, 0)$ is the comoving 4-velocity of the fluid. Then,

$$T_{ab} = \mathrm{diag}(\rho, p, p, p) \tag{23}$$

and the curvature matrix reduces to

$$\mathbf{R}_{AB} = \begin{pmatrix} \mathbf{M}_1 & \mathbf{L} \\ \mathbf{L} & \mathbf{M}_2 \end{pmatrix}, \tag{24}$$

with

$$\mathbf{L} = \begin{pmatrix} \mathbf{R}_{14} & \mathbf{R}_{15} & \mathbf{R}_{16} \\ \mathbf{R}_{15} & \mathbf{R}_{25} & \mathbf{R}_{26} \\ \mathbf{R}_{16} & \mathbf{R}_{26} & -\mathbf{R}_{14} - \mathbf{R}_{25} \end{pmatrix},$$

$$\mathbf{M}_1 = \begin{pmatrix} \mathbf{R}_{11} & \mathbf{R}_{12} & \mathbf{R}_{13} \\ \mathbf{R}_{12} & \mathbf{R}_{22} & \mathbf{R}_{23} \\ \mathbf{R}_{13} & \mathbf{R}_{23} & -\mathbf{R}_{11} - \mathbf{R}_{22} - \Lambda + \frac{\kappa}{2}(3p + \rho) \end{pmatrix},$$

$$\mathbf{M}_2 = \begin{pmatrix} -\mathbf{R}_{11} + \frac{\kappa}{2}(\rho + p) & -\mathbf{R}_{12} & -\mathbf{R}_{13} \\ -\mathbf{R}_{12} & -\mathbf{R}_{22} + \frac{\kappa}{2}(\rho + p) & -\mathbf{R}_{23} \\ -\mathbf{R}_{13} & -\mathbf{R}_{23} & \mathbf{R}_{11} + \mathbf{R}_{22} + \Lambda - \kappa p \end{pmatrix}.$$

Thus, in the case of a perfect fluid solution, the curvature eigenvalues are related by

$$\sum_{i=1}^{6} \lambda_i = \frac{3\kappa}{2}(\rho + p)\,. \tag{25}$$

Finally, in the particular case of vacuum fields, $R_{ab} = 0$, with vanishing cosmological constant, $\Lambda = 0$, the curvature matrix reduces to

$$\mathbf{R}_{AB} = \begin{pmatrix} \mathbf{M} & \mathbf{L} \\ \mathbf{L} & -\mathbf{M} \end{pmatrix}, \tag{26}$$

where

$$
\mathbf{L} = \begin{pmatrix} \mathbf{R}_{14} & \mathbf{R}_{15} & \mathbf{R}_{16} \\ \mathbf{R}_{15} & \mathbf{R}_{25} & \mathbf{R}_{26} \\ \mathbf{R}_{16} & \mathbf{R}_{26} & -\mathbf{R}_{14} - \mathbf{R}_{25} \end{pmatrix}, \qquad \mathbf{M} = \begin{pmatrix} \mathbf{R}_{11} & \mathbf{R}_{12} & \mathbf{R}_{13} \\ \mathbf{R}_{12} & \mathbf{R}_{22} & \mathbf{R}_{23} \\ \mathbf{R}_{13} & \mathbf{R}_{23} & -\mathbf{R}_{11} - \mathbf{R}_{22} \end{pmatrix}, \tag{27}
$$

so that the (3×3) matrices $\mathbf{L}$ and $\mathbf{M}$ are symmetric and trace free,

$$
\mathrm{Tr}(\mathbf{L}) = 0 \,, \qquad \mathrm{Tr}(\mathbf{M}) = 0 \,, \quad \text{i.e.,} \quad \mathrm{Tr}(\mathbf{R}_{AB}) = 0 \,. \tag{28}
$$

Furthermore, the eigenvalues must satisfy the condition

$$
\sum_{i=1}^{6} \lambda_i = 0 \tag{29}
$$

as a consequence of the curvature matrix being traceless.

The explicit form of curvature eigenvalues λ_i depends on the components of the Riemann curvature tensor and behaves as scalars under coordinate transformations. They can, therefore, be used to formulate invariant statements in general relativity. In particular, the properties of λ_i are used to formulate the Petrov classification of gravitational fields [6]. Additionally, the eigenvalue properties have been used to propose an invariant definition of repulsive gravity [9] and alternative cosmological models [10]. Here, we use this idea to propose an invariant formulation of the matching problem in which only curvature eigenvalues are involved.

3. C^3 Matching

The matching between two different spacetimes along a surface Σ is usually performed by using the Darmois and Lichnerowicz conditions (C^2 conditions), which have been shown to be equivalent in a particular coordinate system [4,11–14]. The C^2 conditions state that in certain coordinates, the first fundamental form, i.e., the metrics induced on the matching surface, and the second fundamental form, i.e., the corresponding extrinsic curvatures, must be continuous across Σ. Darmois conditions are represented in a covariant way, which implies that no preference should be given to any particular coordinate system. However, these conditions turn out to be very restrictive in concrete examples, in particular, because the choice of coordinates is a very important step in the sense that the so-called admissible coordinates have to be found in order to apply the matching procedure (for more details see [14]). For instance, in the case of spherically symmetric spacetimes, several options are possible and, therefore, a detailed analysis of each coordinate system should be performed before proceeding with the matching itself [15–18]. One of the advantages of using the C^3 matching procedure is that the results do not depend on the choice of coordinates because we will use only quantities that behave as scalars under a coordinate transformation [9].

Furthermore, an alternative approach was proposed by Israel in [5], which is applied when the extrinsic curvature is not continuous. In fact, in this case, Σ is replaced by a thin shell with an effective energy-momentum tensor, which is defined in terms of the difference of the extrinsic curvature evaluated inside and outside the hypersurface Σ. Since the above matching approaches involve second-order derivatives of the metric, they are known, in general, as C^2 matching.

The C^2 matching is usually difficult to implement because it requires knowing a priori the location of Σ in a particular coordinate system. In the case of compact objects, Σ is identified with the surface of the source of gravity. In general, however, it is quite complicated to find the equation that determines the matching surface, except in cases with a high number of symmetries, such as spherical symmetry, in which the surface is simply a sphere of constant radius.

The main objective of the C^3 procedure is to provide matching conditions that do not depend on the choice of a particular coordinate system and allow us to obtain information about the matching surface Σ. To this end, the C^3 matching uses as a starting point the eigenvalues of the Riemann curvature tensor, which are independent of the choice of coordinate system [6]. In fact, we will also consider the derivatives of the eigenvalues, which involve third-order derivatives of the metric, in order to obtain information about the location of the matching surface. For this reason, we denote our method as C^3 matching.

One of the first applications of the formalism presented above was to formulate an invariant definition of repulsive gravity [9]. The idea of this definition is as follows. In the case of an isolated mass distribution, the corresponding spacetime should be asymptotically flat, and, consequently, all the eigenvalues should vanish at infinity, i.e.,

$$\lim_{r \to \infty} \lambda_i = 0 \; \forall \, i \,, \tag{30}$$

where r is a spatial coordinate that measures the distance to the source of gravity. Then, as the mass distribution is approached, the intensity of the gravitational field should increase, and, correspondingly, the eigenvalues are expected to increase. If an eigenvalue happens to change its sign as the source is approached, we interpret this behavior as an indication of the presence of repulsive gravity. Furthermore, since the eigenvalue vanishes at infinity and increases its value as the object is approached, it should pass through an extremum before changing its sign. To realize this intuitive idea in concrete examples, we proceed as follows. Let the set

$$\{r_l\}, \; l = 1, 2, \ldots \quad \text{with} \quad 0 < r_l < \infty \tag{31}$$

represents the set of solutions to the equation

$$\left. \frac{\partial \lambda_i}{\partial r} \right|_{r=r_l} = 0 \,, \quad \text{with} \quad r_{rep} = \max\{r_l\} \,, \tag{32}$$

i.e., r_{rep} is the location of the first extremum that is found when approaching the source from infinity. We call r_{rep} repulsion radius because at $r = r_{rep}$, the maximum value of attractive gravity is reached, and repulsive gravity starts to play an important role.

The main point now is to use this definition of repulsive gravity in the context of realistic compact objects. In fact, since in the case of compact mass distributions, no repulsive gravity has been detected so far, the idea of the C^3 approach is to replace the region of repulsion ($r < r_{rep}$) with an interior solution of Einstein equations as follows. Indeed, regions of repulsive gravity have been shown to exist in Reissner–Nordström, Kerr, and Kerr–Newman black holes [9] as well as in gravitational fields generated by a mass distribution with quadrupole moment [19]. In such cases, the matching with an interior solution should be performed in such a way that the matching surface is located outside the region of repulsive gravity.

Let us consider an exterior spacetime $(M^+, g^+_{\mu\nu})$ and an interior spacetime $(M^-, g^-_{\mu\nu})$ with curvature eigenvalues $\{\lambda_i^+\}$ and $\{\lambda_i^-\}$, respectively. Then, the C^3 matching approach consists of two steps:

(*i*)   Define the matching surface Σ by means of the matching radius r_{match}, defined as

$$r_{match} \in [r_{rep}, \infty) \,, \quad \text{with} \quad r_{rep} = \max\{r_l\} \,, \quad \left. \frac{\partial \lambda_i^+}{\partial r} \right|_{r=r_l} = 0 \,. \tag{33}$$

This means that the repulsion radius is determined by the location of the first extremum that is found when approaching the source of gravity from infinity.

(*ii*)   Perform the matching of the spacetimes $(M^+, g^+_{\mu\nu})$ and $(M^-, g^-_{\mu\nu})$ at Σ by imposing the conditions

$$\left. \lambda_i^+ \right|_{\Sigma} = \left. \lambda_i^- \right|_{\Sigma} \quad \forall i \,. \tag{34}$$

In other words, the C^3 matching consists in demanding that the curvature eigenvalues be continuous across the matching surface Σ, which should be located anywhere between the repulsion radius and infinity. Thus, the idea of the C^3 matching is to avoid the presence of repulsive gravity in the case of gravitational compact objects. Conditions (33) and (34) turn out to be very restrictive in the sense that they do not allow the case of discontinuities across the matching surface. We will see in Section 4 that the C^3 matching procedure can be generalized to include this case too.

4. The Spherically Symmetric Matching

In the case of spherically symmetric gravitational fields, the exterior spacetime is unique and is described by the Schwarzschild line element

$$ds^2 = -\left(1 - \frac{2M}{r}\right)dt^2 + \frac{dr^2}{1 - \frac{2M}{r}} + r^2(d\theta^2 + \sin^2\theta d\varphi^2)\,, \tag{35}$$

where M represents the mass of the gravitational source. The orthonormal tetrad can be chosen in the canonical form

$$\vartheta^0 = \left(1 - \frac{2M}{r}\right)^{1/2}dt\,, \quad \vartheta^1 = \left(1 - \frac{2M}{r}\right)^{-1/2}dr\,, \quad \vartheta^2 = rd\theta\,, \quad \vartheta^3 = r\sin\theta d\varphi. \tag{36}$$

A straightforward computation shows that, in this case, the curvature matrix has the form

$$\mathbf{R}_{AB} = \begin{pmatrix} -\frac{2M}{r^3} & 0 & 0 & 0 & 0 & 0 \\ 0 & \frac{M}{r^3} & 0 & 0 & 0 & 0 \\ 0 & 0 & \frac{M}{r^3} & 0 & 0 & 0 \\ 0 & 0 & 0 & \frac{2M}{r^3} & 0 & 0 \\ 0 & 0 & 0 & 0 & -\frac{M}{r^3} & 0 \\ 0 & 0 & 0 & 0 & 0 & -\frac{M}{r^3} \end{pmatrix}, \tag{37}$$

Then, the eigenvalues are determined by the diagonal elements of the matrix $\mathbf{R}_{AB}$ and we obtain

$$\lambda_1^+ = -\lambda_4^+ = -\frac{2M}{r^3}, \ \lambda_2^+ = \lambda_3^+ = -\lambda_5^+ = -\lambda_6^+ = \frac{M}{r^3}. \tag{38}$$

For the investigation of the interior spacetime M^-, we consider the general spherically symmetric line element

$$ds^2 = -e^\nu dt^2 + e^\phi dr^2 + r^2(d\theta^2 + \sin^2\theta d\varphi^2) \tag{39}$$

where ν and ϕ are functions that depend on r only. It then follows that the orthonormal tetrad can be chosen as

$$\vartheta^0 = e^{\nu/2}dt\,, \quad \vartheta^1 = e^{\phi/2}dr\,, \quad \vartheta^2 = rd\theta\,, \quad \vartheta^3 = r\sin\theta d\varphi\,. \tag{40}$$

Using Cartan's structure equations, we obtain the following non-vanishing components of the curvature matrix:

$$\mathbf{R}_{11} = -\frac{1}{4}(\phi_{,r}\nu_{,r} - \nu_{,r}^2 - 2\nu_{,rr})e^{-\phi}, \ \mathbf{R}_{22} = \frac{1}{2r}\nu_{,r}e^{-\phi}, \ \mathbf{R}_{33} = \frac{1}{2r}\nu_{,r}e^{-\phi}, \tag{41}$$

$$\mathbf{R}_{44} = \frac{1}{r^2}(1 - e^{-\phi}), \ \mathbf{R}_{55} = \frac{1}{2r}\phi_{,r}e^{-\phi}, \ \mathbf{R}_{66} = \frac{1}{2r}\phi_{,r}e^{-\phi}, \tag{42}$$

where we have used Einstein's equations in the form

$$\nu_{,rr} + \frac{1}{2}\nu_{,r}^2 - \frac{\nu_{,r}}{2r}(2 + r\phi_{,r}) - \frac{\phi_{,r}}{r} - \frac{2}{r^2}(1 - e^\phi) = 0\,, \tag{43}$$

$$\kappa\rho = \frac{1}{r^2}[1 + e^{-\phi}(r\phi_{,r} - 1)], \ \kappa p = -\frac{1}{r^2}[1 - e^{-\phi}(1 + r\nu_{,r})]. \tag{44}$$

Then, we obtain the following eigenvalues for the curvature tensor of a spherically symmetric interior perfect fluid solution

$$\lambda_1^- = -\frac{1}{4}(\phi_{,r}v_{,r} - v_{,r}^2 - 2v_{,rr})e^{-\phi}, \tag{45}$$

$$\lambda_2^- = \lambda_3^- = \frac{1}{2r}v_{,r}e^{-\phi}, \tag{46}$$

$$\lambda_4^- = \frac{1}{4}(\phi_{,r}v_{,r} - v_{,r}^2 - 2v_{,rr})e^{-\phi} + \frac{k(\rho + p)}{2}, \tag{47}$$

$$\lambda_5^- = \lambda_6^- = -\frac{1}{2r}v_{,r}e^{-\phi} + \frac{k(\rho + p)}{2}. \tag{48}$$

The computation of the C^3 matching condition $d\lambda_i^+/dr = 0$ shows that there is no repulsion radius, implying that the matching can be carried out within the interval $r_{match} \in (0, \infty)$. The second matching condition implies that the exterior (38) and interior eigenvalues (45) coincide on the matching surface. This leads to the following set of independent equations

$$-\frac{1}{4}(\phi_{,r}v_{,r} - v_{,r}^2 - 2v_{,rr})e^{-\phi}, = -\frac{2M}{r^3} \tag{49}$$

$$\frac{1}{2r}v_{,r}e^{-\phi} = \frac{M}{r^3}, \tag{50}$$

$$\frac{1}{4}(\phi_{,r}v_{,r} - v_{,r}^2 - 2v_{,rr})e^{-\phi} + \frac{k(\rho + p)}{2} = \frac{2M}{r^3}, \tag{51}$$

$$-\frac{1}{2r}v_{,r}e^{-\phi} + \frac{k(\rho + p)}{2} = -\frac{M}{r^3}. \tag{52}$$

The above system of algebraic equations has to be satisfied in order for an arbitrary perfect fluid solution to be matched with the exterior Schwarzschild spacetime. It is easy to show that the above set of algebraic conditions allows only one solution, namely,

$$\rho = 0, \ p = 0. \tag{53}$$

This result corroborates in an invariant way our physical expectation of vanishing pressure and density on the matching surface. This result contrasts with the one obtained by using the Darmois matching conditions, according to which perfect-fluid interior solutions with non-zero densities and pressures at the matching surface, described by a sphere of constant radius, are configurations that can be matched with the exterior Schwarzschild spacetime [8]. In this sense, the Israel matching conditions offer an additional possibility, according to which the non-zero values of the density and pressure on the matching surface are due to the presence of a thin shell with exactly those values of density and pressure. In the resulting configuration, the matching problem is transferred to the thin shell, which is described by an energy-momentum tensor whose physical meaning has to be established separately [5,8].

In the C^3 approach, it is also possible to generalize the matching conditions to include non-zero values of the energy density and pressure. Indeed, as shown in [8], a discontinuous matching can be performed explicitly if we assume that the Einstein tensor $G_{ij}^{\pm}$ induced on the matching surface Σ satisfies the condition

$$G_{ij}^- - G_{ij}^+ = kS_{ij} \tag{54}$$

where k is a real constant and S_{ij} is a well-defined energy-momentum tensor. This means that Einstein's equations are satisfied across the matching surface. In the case of perfect-fluid solutions, the tensor S_{ij} is physically relevant if it is defined as

$$S_{ij} = T_{ij}^- - T_{ij}^+ = [(\sigma + P)u_i u_j + P\gamma_{ij}], \tag{55}$$

where $\sigma = \rho|_\Sigma$ and $P = p|_\Sigma$ are the non-zero values of the density and pressure at the matching surface, respectively, and γ_{ij} is the spacetime metric induced on Σ. Thus, we see that the non-zero values of the density and pressure at the matching surface can be used to construct an energy-momentum tensor that guarantees the fulfillment of the Einstein equations on the matching conditions so that the matching can be performed explicitly. Several examples of the application of this procedure have been presented in [8]. In particular, the case in which the surface pressure P vanishes can be represented as

$$S_{ij} = 2(\lambda_1^- - \lambda_1^+)u_i u_j, \tag{56}$$

indicating that the surface density σ can be represented invariantly in terms of the eigenvalues. This particular result could be used to apply the matching procedure in the case of strange stars [20].

5. Final Remarks and Perspectives

In this work, we presented an invariant formalism to apply matching conditions in general relativity, which is based upon the use of the eigenvalues of the Riemann curvature tensor and its derivatives. In this C^3 approach, we demand that the curvature eigenvalues of the exterior and interior solutions be continuous across the matching surface. In addition, the derivatives of the eigenvalues are used to determine the location of the matching surface. In this work, we limit ourselves to the case of isolated gravitational sources so that the curvature and the eigenvalues vanish at spatial infinity. Then, we look at the behavior of the eigenvalues as the source of gravity is approached from infinity. We argue that if an eigenvalue shows local extrema and changes its sign as the source is approached, this is an effect due to the presence of repulsive gravity. In fact, this behavior has been used to propose an invariant definition of repulsive gravity, which includes the concept of the radius of repulsion as corresponding to the location of the first extremum that appears as the source is approached from spatial infinity. Furthermore, we define the matching radius as the minimum radius where the matching can be performed. In other words, the matching surface can be located anywhere between the location of the repulsion radius and infinity. The goal of fixing a minimum radius for the matching surface is to avoid the presence of repulsive gravity because, so far, it has not been detected in the gravitational field of compact astrophysical objects.

We analyze in detail the case of a spherically symmetric mass distribution, in which the exterior field is described by the Schwarzschild spacetime, and the interior counterpart corresponds to a perfect fluid. It is interesting to note that due to the versatility of the C^3 matching formalism in the sense that the curvature eigenvalues can be calculated in general for any metric without specifying any particular solution, it is not necessary to fix the interior perfect-fluid solution. We use instead the general form of the matrix curvature that satisfies Einstein's equations. First, we notice that the derivatives of the exterior eigenvalues do not have any extrema, a result that we interpret as indicating that there is no repulsion radius and the matching can be performed at any place between the origin of coordinates and spatial infinity. Then, we find the set of algebraic equations that follows from the condition that the interior and exterior eigenvalues coincide at the matching surface. It turns out that this set of equations allows only one solution, namely, that the pressure and density should vanish at the matching surface. We conclude that the C^3 matching procedure in the case of a spherically symmetric gravitational field leads to the results expected from a physical point of view.

Another case of interest is that of stationary axially symmetric fields, which allows the analysis of rotating gravitational fields. In the case of vacuum, the general line element can be written in cylindrical coordinates (t, ρ, z, φ) as [6]

$$ds^2 = e^{2\psi}(dt - \omega d\varphi)^2 - e^{-2\psi}\left[e^{2\gamma}(d\rho^2 + dz^2) + \rho^2 d\varphi^2\right], \tag{57}$$

where ψ, ω, and γ are functions of ρ and z, only. The calculation of the corresponding curvature eigenvalues and their derivatives with respect to ρ and z leads to a set of equations that should determine the location of the matching surface. However, it is not easy to interpret the significance of the results, probably because it is necessary to use a different set of coordinates. We expect to investigate this problem in future works. In the particular case of a slowly rotating mass, whose gravitational field can be described by the exterior Lense–Thirring metric and the interior Hartle–Thorne approximate solution [6], the matching conditions lead to a system of equations that must be solved numerically. Work in this direction is in progress.

The particular case of static axially symmetric gravitational ($\omega = 0$) is interesting because it resembles the case of Newtonian gravity. Indeed, in this case, the field equation that determines the function ψ turns out to be linear, and its general asymptotically flat solution can be written as [21]

$$\psi = \sum_{n=0}^{\infty} \frac{a_n}{(\rho^2 + z^2)^{\frac{n+1}{2}}} P_n(\cos\theta), \qquad \cos\theta = \frac{z}{\sqrt{\rho^2 + z^2}}, \tag{58}$$

where a_n ($n = 0, 1, \ldots$) are arbitrary constants, and $P_n(\cos\theta)$ represents the Legendre polynomials of degree n. The solution for the function γ can be obtained from the above expression by quadratures. Interestingly, the solution (58) coincides with the exterior Newtonian potential given in Equation (7). This coincidence could be used to search for an interior line element, in which the function corresponding to ψ could be given as an infinite series in terms of the Green function (5). This has been done in the particular case of a metric with a quadrupole moment in [22,23]. We plan to continue the study of this problem in future works.

Another interesting aspect that has not been explored in the C^3 formalism is the possibility of analyzing the internal structure of compact objects by using thin shells, especially regarding stability properties and phase transition structures [24]. To this end, it will be necessary to investigate the dynamics of thin shells determined by the matching conditions in the presence of discontinuities as given in Equations (54)–(56). This is an interesting open question that deserves further development.

Funding: This research was funded by UNAM-DGAPA-PAPIIT, Grant No. 114520, and CONACYT-Mexico, Grant No. A1-S-31269.

Acknowledgments: It is an honor to dedicate this work to Remo Ruffini on the occasion of his 80th anniversary.

Conflicts of Interest: The author declares no conflict of interest.

References

1. Misner, C.W.; Thorne, K.S.; Wheeler, J.A. *Gravitation*; Freeman: San Francisco, CA, USA, 2000.
2. Schwarzschild, K. Über das Gravitationsfeld eines Massenpunktes nach der Einsteinschen Theorie. *Sitzungsber. Preuss. Akad. Wissensch.* **1916**, *18*, 189.
3. Schwarzschild, K. Über das Gravitationsfeld einer Kugel aus inkompressibler Flüssigkeit nach der Einsteinschen Theorie. *Sitz. Deut. Akad. Wiss. Math.-Phys.* **1916**, *24*, 424.
4. Darmois, G. *Les équations de la Gravitation Einsteinienne*; Gauthier-Villars: Paris, France, 1927.
5. Israel, W. Singular hypersurfaces and thin shells in general relativity. *Il Nuovo Cimento B* **1966**, *44*, 1. [CrossRef]
6. Stephani, H.; Kramer, D.; MacCallum, M.; Hoenselaers, C.; Herlt, E. *Exact Solutions of Einstein's Field Equations*; Cambridge University Press: Cambridge, UK, 2009.

7. Gutierrez-Piñerez, A.; Quevedo, H. C3 matching for asymptotically flat spacetimes. *Class. Quantum Grav.* **2019**, *36*, 135003. [CrossRef]
8. Gutierrez-Piñerez, A.; Quevedo, H. Darmois matching and C3 matching. *Class. Quantum Grav.* **2022**, *39*, 035015. [CrossRef]
9. Luongo, O.; Quevedo, H. Characterizing repulsive gravity with curvature eigenvalues. *Phys. Rev. D* **2014**, *90*, 084032. [CrossRef]
10. Luongo, O.; Quevedo, H. Self-accelerated universe induced by repulsive effects as an alternative to dark energy and modified gravities. *Found. Phys.* **2018**, *48*, 17. [CrossRef]
11. Robson, E.H. Junction conditions in general relativity theory. *Annales de l'Institut Henri Poincaré* **1972**, *16*, 41.
12. Bonnor, W.B.; Vickers, P.A. Junction conditions in general relativity. *Gen. Relativ. Gravit.* **1981**, *13*, 29. [CrossRef]
13. Raju, C.K. Junction conditions in general relativity. *J. Phys. A Math. Gen.* **1982**, *15*, 1785. [CrossRef]
14. Lake, K. Revisiting the Darmois and Lichnerowicz junction conditions. *Gen. Relativ. Gravit.* **2017**, *49*, 134. [CrossRef]
15. Hernández-Pastora, J.L.; Martín, I.; Ruiz, E. Admissible coordinates of Lichnerowicz for the Schwarzschild metric. *arXiv* **2001**, arXiv:gr-qc/0109031v3.
16. Leibovitz, C. Junction Conditions for Spherically Symmetric Matter in Co-moving Co-ordinates. *Il Nuovo Cimento* **1969**, *60*, 254. [CrossRef]
17. Shen, W.; Zhu, S. Junction conditions on null hypersurface. *Phys. Lett. A* **1988**, *126*, 229.
18. Dandach, N.F. Matching of Gravitational Fields in General Relativity: Junction Conditions in Synchronous and in Gaussian Coordinates. *Il Nuovo Cimento B* **1992**, *107*, 1267. [CrossRef]
19. Quevedo, H. Mass quadrupole as a source of naked singularities. *Int. J. Mod. Phys. D* **2011**, *20*, 1779. [CrossRef]
20. Alcock, C.; Farhi, E.; Olinto, A. Strange stars. *Astrophys. J.* **1986**, *310*, 261. [CrossRef]
21. Weyl, H. Zur Gravitationstheorie. *Ann. Phys.* **1917**, *54*, 117. [CrossRef]
22. Toktarbay S.; Quevedo, H. A stationary q-metric. *Grav. Cosm.* **2014**, *20*, 252. [CrossRef]
23. Toktarbay, S.; Quevedo, H.; Abishev, M.; Muratkhan, A. Gravitational field of slightly deformed naked singularities. *Eur. Phys. J. C* **2022**, *82*, 382. [CrossRef]
24. Pereira, J.P.; Coelho, J.G.; Rueda, J.A. Stability of thin-shell interfaces inside compact stars. *Phys. Rev. D* **2014**, *90*, 123011. [CrossRef]

Communication

Binary Neutron-Star Mergers with a Crossover Transition to Quark Matter

Grant J. Mathews [1,*], Atul Kedia [1], Hee Il Kim [2] and In-Saeng Suh [1,3,4]

[1] Department of Physics and Astronomy, Center for Astrophysics, University of Notre Dame, Notre Dame, IN 46556, USA; atulkedia93@gmail.com (A.K.); suhi@ornl.gov or isuh@nd.edu.it (I.-S.S.)
[2] Center for Quantum Spacetime, Sogang University, Seoul 04107, Republic of Korea
[3] Center for Research Computing, University of Notre Dame, Notre Dame, IN 46556, USA
[4] National Center for Computational Sciences, Oak Ridge National Laboratory, Oak Ridge, TN 37830, USA
* Correspondence: gmathews@nd.edu

Abstract: This paper summarizes recent work on the possible gravitational-wave signal from binary neutron-star mergers in which there is a crossover transition to quark matter. Although this is a small piece of a much more complicated problem, we discuss how the power spectral density function may reveal the presence of a crossover transition to quark matter.

Keywords: neutron stars; equation of state; quark matter; gravitational waves

Citation: Mathews, G.J.; Kedia, A.; Kim, H.I.; Suh, I.-S. Binary Neutron-Star Mergers with a Crossover Transition to Quark Matter. *Universe* **2023**, *9*, 410. https://doi.org/10.3390/universe9090410

Academic Editors: Nicolas Chamel, Jorge Armando Rueda Hernández, Narek Sahakyan and Gregory Vereshchagin

Received: 7 July 2023
Revised: 1 August 2023
Accepted: 30 August 2023
Published: 7 September 2023

1. Introduction

I am honored to have a chance to contribute to this Festschrift in honor of Remo Ruffini's 80th birthday. I have enjoyed collaboration with Remo on papers [1,2] exploring the physics of the X-ray afterglow associated gamma-ray bursts. Indeed, this collaboration inspired me to return to simulations of the relativistic hydrodynamics associated with binary neutron stars and their merger. As Remo has correctly pointed out during this Festschrift and elsewhere, the electromagnetic evolution will dominate the dynamics of binary neutron-star mergers, and moreover, there are enormous uncertainties associated with detecting and calculating the gravitational radiation emanating from binary neutron-star mergers. Nevertheless, in this presentation, I describe recent work [3] with my collaborators in which we have simulated the relativistic merger of neutron stars and explored effects on the emergent gravitational waves of a crossover transition to quark matter.

It has been discussed for some time that neutron stars (NSs) within binary systems could be used to probe the equation of state (EoS) at high densities (e.g., Refs. [4,5]). Gravitational waves (GWs) from the GW170817 event by the LIGO-Virgo Collaboration [6,7] may have provided new insight into neutron-star matter [8]. Also, NS masses and radii determined by the NICER mission constrain the EoS of nuclear matter [9–11].

Moreover, differences in the EoS can lead to a variety of observable effects (cf. [12]). Such changes in the EoS may lead to a change in the maximum peak frequency f_{peak} (sometimes denoted as f_2) in the inferred power spectral density (PSD) [13–15]. A shift may violate the proposed universality relations between f_{peak} and tidal deformability for pure hadronic EoSs [16–21]. In [3], we analyzed how such an observed shift might also probe the quark matter phase. This is, however, model dependent (e.g., [22,23]) and depends somewhat on the time duration of the merged system [12,24,25].

There have been many recent works considering EoS effects on the detected GWs. Some have considered the formation of quark matter [12–14,23–33]. Often these studies, however, were limited to a first-order phase transition. In a first order transition, a mixed phase of quarks and hadrons develops. This mixed phase diminishes the pressure support of the remnant, resulting in a prompt collapse. However, a crossover or a weak first-order transition remains a possibility [34–38]. The matter pressure during the crossover could be enormous. This could extend the duration of the postmerger phase. We proposed that observing such a long-duration post-merger system might signal both the phase-transition order and the strength of the coupling of quark matter [3].

In Ref. [3], we calculated the GW signal from the postmerger phase and showed that it is sensitive to the presence of quark matter in the equation of state. We demonstrated that the properties of quark matter in the crossover phase increase the duration of the postmerger GW emission. Hence, this probes the properties of quark matter. Various parameterizations of the quark-hadron crossover (QHC19) EoS of [39] were investigated in Ref. [3] . A similar study was made [40] based on the more recent version of the (QHC21) EoS with similar conclusions. The crossover is treated as continuous in the QHC19 EoS. In Ref. [3], the maximum chirp frequency f_{max}, the tidal deformability, and the peak in the power spectral density f_{peak} were used to identify observational characteristics of a crossover to quark matter during mergers of equal-mass binary neutron stars. The crucial high frequency range (1–4 kHz) is associated with the postmerger gravitational waves. Although this frequency is not within the sensitivity limits of the LIGO/aVirgo/KAGRA observatories, the next generation of gravitational-wave detectors, e.g., the Einstein Telescope [41] and Cosmic Explorer [42], should be sensitive to such high frequency emissions. We argue that this observation in the next generation detectors might indicate both the order of the transition and the parameters characterizing the crossover to quark matter.

2. Equations of State

At high baryon density a non-perturbative approach to QCD is required. This approach must include chiral symmetry breaking [43], the generation of constituent quark masses, quark pairing, the possibility of color superconductivity [44], etc. In the hadronic regime, we considered both the SLy [45] and the GNH3 [46] EoSs. These bracket the properties of an extremely soft or a rather stiff equation of state.

The QHC19 EoS is based upon the the NJL Lagrangian [47–49]. The four coupling constants are: (1) (G) the scalar coupling; (2) (K) the coefficient of the Kobayashi-Maskawa-'t Hooft vertex; (3) (g_v) the vector coupling for quark repulsion; and (4) (H) the di-quark strength. Only two coupling constants $(g_v/G$ and $H/G)$, were used to construct various versions of the EoS. The three parameter sets used in Ref. [3] are labeled as [39]: QHC19B $[(g_V, H) = (0.8, 1.49)]$, QHC19C $[(g_V, H) = (1.0, 1.55)]$, and QHC19D $[(g_V, H) = (1.2, 1.61)]$. At the crossover densities ($2\, n_0 < n < 5\, n_0$), the pressure is given by fifth-order polynomials in terms of the baryonic chemical potential.

3. Simulation Details

Binary merger simulations were run in [3] using the `Einstein Toolkit` [50] numerical relativity software. This includes full general relativity in three spatial dimensions with differential equations based upon the BSSN-NOK framework [51–55]. The hydrodynamics was evolved with the use of the `GRHydro` code [56–58] based on the Valencia formulation [59,60]. The initial conditions were generated using `LORENE` [61,62]. The thorn `Carpet` [63,64] was used for adaptive mesh refinement based upon six mesh refinement levels and a minimum grid of 0.3125 in Cactus units ($\approx$461 m). A constant adiabatic index $\Gamma_{th} = 1.8$ was used to account for the thermal pressure in `GRHydro` as described in Ref. [65].

The Newman–Penrose formalism was employed to extract the gravitational waves emitted during the binary merger. This minimizes numerical noise by fitting a multipole expansion in spherical harmonics of the Weyl scalar $\Psi_4^{(l,m)}(\theta, \phi, t) = \ddot{h}_+^{(l,m)}(\theta, \phi, t) + i\ddot{h}_\times^{(l,m)}(\theta, \phi, t)$. The two polarizations of the strain $h_+(\theta, \phi, t)$ and $h_\times(\theta, \phi, t)$ result from a sum over the (l, m) modes followed by integrating twice. The neutron star models were based upon baryonic masses of $M_B = 1.45, 1.50, 1.55\ M_\odot$. These were chosen because gravitational masses associated with these baryonic masses for various equations of state are similar $\sim$1.35–1.4 $M_\odot$. Simulations began at an initial coordinate separation of 45 km between centers.

In Ref. [3] it was shown that even during inspiral, the central densities in the neutron stars achieved densities in the crossover range (2–5 n_0). During the merger, the maximum density increases until it exceeds $\sim$5–6 n_0. The central region of the system then collapses.

It was also shown in Ref. [3] that the postmerger GW emission continues for a much longer time for the simulations with a QHC EoS. When going from QHCB to QHCC the postmerger GW emission becomes longer, corresponding to increasing the quark coupling. This led to the suggestion that the strength of the quark–matter couplings might be deduced from the duration of the post merger phase. Indeed, the lifetime of the postmerger intermediate hyper-massive neutron star (HMNS) depends rather significantly on the stiffness of the equation of state at the crossover densities. One interesting finding is that the postmerger remnants from mergers including the QHC19D EoS had so much pressure that no black hole formed during the simulations. As the EoS stiffness within the QHC models increased lifetimes of their HMNS remnants were apparent. Even the QHC19B EoS produces a much longer postmerger duration than that of a pure hadronic EoSs.

A waveform analysis of the strain can be performed in the frequency domain. This highlights the dominant frequencies of the waveform. Specifically, the effective Fourier amplitude is obtained from

$$\tilde{h}_{+,\times}(f) = \int h_{+,\times}(t) e^{-i2\pi ft} dt \ . \tag{1}$$

This is presented in Figure 1, which shows an example of the normalized power spectral density $2\tilde{h}(f)f^{1/2}$ [66] based upon the simulations of Ref. [3]. The lower blue and orange curves show the anticipated sensitivity of the Einstein Telescope and Cosmic Explorer, respectively, while the upper green curve shows the current LIGO sensitivity. The initial inspiral up to contact between the merging neutron stars ends with the first peak at around 1 kHz. For probing the crossover to quark matter, however, the peaks, f_{peak}, at around 2.5 3.5 kHz are most useful. These arise from the extended postmerger phase. The amplitude of f_{peak} correlates with the time duration of the postmerger remnant. Therefore, it correlates with the strength of the coupling constants in the QHC19 equations of state. As discussed in [3], one can also infer the maximum chirp strain amplitude, $f_{max} = \frac{1}{2\pi}\frac{d\phi}{dt}|_{max}$, where ϕ is the phase of the strain (see [66]). This is not apparent in the PSD, but is deduced from the phase of the strain during the merger. Although this is referred to at the maximum chirp strain, this is not to be confused with the instantaneous gravitational-wave frequency at the time of merger.

4. Discussion and Conclusions

Although the amplitude of the f_{peak} PSD becomes larger for crossover equations of state with increasing coupling strengths, this might also be realized in other equations of state as demonstrated in [66]. Therefore, one desires another signature to uniquely show the formation of quark matter. In [3], it was pointed out that the QHC19 equations of state show behavior consistent with a soft EoS at low density, $\sim$3n_0. This affects the merger regime of f_{max}. On the other hand, the postmerger phase represented in the f_{peak} frequency exhibits the behavior of a stiff EoS.

This dual nature of the QHC19 EoSs might be revealed by correlating f_{max} and f_{peak} in a GW event [3]. This is illustrated in Figure 2. For pure hadronic EoSs, there appears to be a linear correlation between f_{max} and f_{peak}. However, a crossover EoS deviates from this correlation as indicated by the circled points on this figure. Thus, an observation of events in the circled region might indicate the crossover to quark matter. We note, however, that this deviation is not entirely robust as an indicator. For example, the hadronic Sly EoS also deviates from the linear relation. What is needed is a more exhaustive set of calculations to better clarify this trend. That, however, is left to a future work.

Additionally, in Ref. [3], the relation between f_{peak} and the pseudo-averaged rest-mass density [17,66] was considered. For this case, the f_{peak} frequencies tend to cluster in a region in between a soft and stiff EoS [3]. Hence, although there are enormous uncertainties in this suggestion, observing a transition from soft to stiffness in the correlations of f_{max} and f_{peak} could indicate that quark matter had formed during the merger. Moreover, the amplitude of the PSD at the frequency of f_{peak} may suggest the quark–matter coupling strengths.

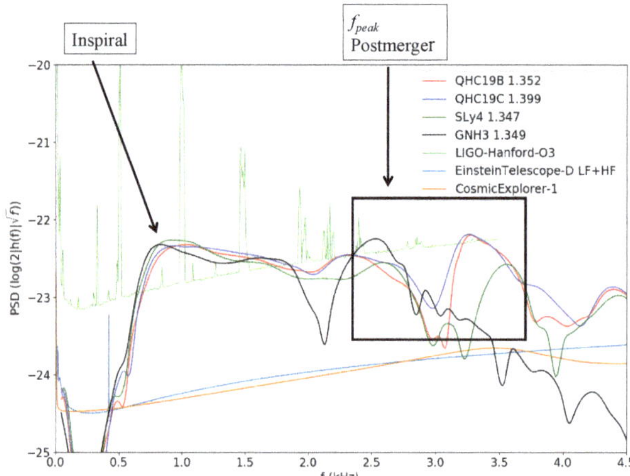

Figure 1. Power spectral density $(2\tilde{h}(f)f^{1/2})$ vs. frequency f for various simulations. The the lower blue and orange curves show anticipated sensitivity of the Einstein Telescope and Cosmic Explorer, respectively, while the upper green curve shows the LIGO sensitivity. The first peak at around 1 kHz is the initial contact of the merging binaries. The second peaks near 2.5–3.5 kHz correspond to the long postmerger phase, f_{peak}.

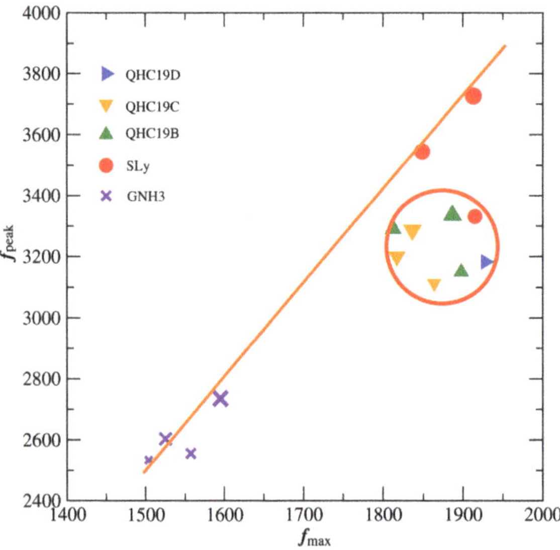

Figure 2. Correlation between f_{max} and f_{peak}. There appears to be a linear correlation for normal hadronic EoSs as indicated by the straight line. However, the existence of a crossover regime to quark matter leads to outliers from this correlation as indicated by the circled points.

Author Contributions: Conceptualization, H.I.K., I.-S.S., A.K. and G.J.M. methodology, H.I.K. and A.K.; software, H.I.K., I.-S.S. and A.K.; validation, A.K., I.-S.S. and H.I.K.; formal analysis, H.I.K., A.K., G.J.M. and I.-S.S.; investigation, A.K., H.I.K., I.-S.S. and G.J.M.; resources, H.I.K., I.-S.S., G.J.M.; data curation, A.K., H.I.K. and I.-S.S. writing—original draft preparation, G.J.M.; writing—review and editing, G.J.M., I.-S.S. and A.K.; visualization, A.K. and H.I.K.; supervision, G.J.M. and H.I.K.; project administration, G.J.M. and H.I.K.; funding acquisition, G.J.M. and H.I.K. All authors have read and agreed to the published version of the manuscript.

Funding: Work at the Center for Astrophysics of the University of Notre Dame is supported by the U.S. Department of Energy under Nuclear Theory Grant No. DE-FG02-95-ER40934. This research was supported in part by the Notre Dame Center for Research Computing through the high performance computing resources. The work of H.I.K. was supported by Basic Science Research Program through the National Research Foundation of Korea (NRF) funded by the Ministry of Education through the Center for Quantum Spacetime (CQUeST) of Sogang University (Grant No. NRF-2020R1A6A1A03047877). This research used resources of the Oak Ridge Leadership Computing Facility at the Oak Ridge National Laboratory, which is supported by the Office of Science of the U.S. Department of Energy under Contract No. DE-AC05-00OR22725.

Data Availability Statement: The DOE will provide public access to these results in accordance with the DOE Public Access Plan (http://energy.gov/downloads/doe-public-access-plan, accessed on 28 August 2023).

Acknowledgments: H.I.K. graciously thanks Jinho Kim and Chunglee Kim for continuous support. This manuscript has in part been authored by UT-Battelle, LLC under Contract No. DE-AC05-00OR22725 with the U.S. Department of Energy. The publisher, by accepting the article for publication, acknowledges that the U.S. Government retains a non-exclusive, paid up, irrevocable, world-wide license to publish or reproduce the published form of the manuscript, or allow others to do so, for U.S. Government purposes. Software used was as follows The Einstein Toolkit (Ref. [50]; https://einsteintoolkit.org, accessed on 28 August 2023), LORENE (Refs. [61,62]), PyCactus (https://bitbucket.org/GravityPR/pycactus, accessed on 28 August 2023), and TOVsolver (https://github.com/amotornenko/TOVsolver, accessed on 28 August 2023).

Conflicts of Interest: The authors declare no conflict of interest.

References

1. Ruffini, R.; Karlica, M.; Sahakyan, N.; Rueda, J.A.; Wang, Y.; Mathews, G.J.; Bianco, C.L.; Muccino, M. A GRB Afterglow Model Consistent with Hypernova Observations. *Astrophys. J.* **2018**, *869*, 101. [CrossRef]
2. Ruffini, R.; Moradi, R.; Rueda, J.A.; Li, L.; Sahakyan, N.; Chen, Y.C.; Wang, Y.; Aimuratov, Y.; Becerra, L.; Bianco, C.L.; et al. The morphology of the X-ray afterglows and of the jetted GeV emission in long GRBs. *MNRAS* **2021**, *504*, 5301–5326. [CrossRef]
3. Kedia, A.; Kim, H.I.; Suh, I.S.; Mathews, G.J. Binary neutron star mergers as a probe of quark-hadron crossover equations of state. *Phys. Rev. D* **2022**, *106*, 103027. [CrossRef]
4. Baiotti, L. Gravitational waves from neutron star mergers and their relation to the nuclear equation of state. *Prog. Part. Nucl. Phys.* **2019**, *109*, 103714. [CrossRef]
5. Radice, D.; Bernuzzi, S.; Perego, A. The Dynamics of Binary Neutron Star Mergers and GW170817. *Annu. Rev. Nucl. Part. Sci.* **2020**, *70*, 95–119. [CrossRef]
6. Abbott, B.P.; Abbott, R.; Abbott, T.D.; Acernese, F.; Ackley, K.; Adams, C.; Adams, T.; Addesso, P.; Adhikari, R.X.; Adya, V.B.; et al. GW170817: Observation of Gravitational Waves from a Binary Neutron Star Inspiral. *Phys. Rev. Lett.* **2017**, *119*, 161101. [CrossRef]
7. Abbott, B.P.; Abbott, R.; Abbott, T.D.; Acernese, F.; Ackley, K.; Adams, C.; Adams, T.; Addesso, P.; Adhikari, R.X.; Adya, V.B.; et al. GW170817: Measurements of Neutron Star Radii and Equation of State. *Phys. Rev. Lett.* **2018**, *121*, 161101. [CrossRef] [PubMed]
8. Lattimer, J.M. The Nuclear Equation of State and Neutron Star Masses. *Annu. Rev. Nucl. Part. Sci.* **2012**, *62*, 485–515. [CrossRef]
9. Bogdanov, S.; Guillot, S.; Ray, P.S.; Wolff, M.T.; Chakrabarty, D.; Ho, W.C.G.; Kerr, M.; Lamb, F.K.; Lommen, A.; Ludlam, R.M.; et al. Constraining the Neutron Star Mass-Radius Relation and Dense Matter Equation of State with NICER. I. The Millisecond Pulsar X-Ray Data Set. *Astrophys. J. Lett.* **2019**, *887*, L25. [CrossRef]
10. Bogdanov, S.; Dittmann, A.J.; Ho, W.C.G.; Lamb, F.K.; Mahmoodifar, S.; Miller, M.C.; Morsink, S.M.; Riley, T.E.; Strohmayer, T.E.; Watts, A.L.; et al. Constraining the Neutron Star Mass-Radius Relation and Dense Matter Equation of State with NICER. III. Model Description and Verification of Parameter Estimation Codes. *Astrophys. J. Lett.* **2021**, *914*, L15. [CrossRef]
11. Miller, M.C.; Lamb, F.K.; Dittmann, A.J.; Bogdanov, S.; Arzoumanian, Z.; Gendreau, K.C.; Guillot, S.; Ho, W.C.G.; Lattimer, J.M.; Loewenstein, M.; et al. The Radius of PSR J0740+6620 from NICER and XMM-Newton Data. *Astrophys. J. Lett.* **2021**, *918*, L28. [CrossRef]

12. Weih, L.R.; Hanauske, M.; Rezzolla, L. Postmerger Gravitational-Wave Signatures of Phase Transitions in Binary Mergers. *Phys. Rev. Lett.* **2020**, *124*, 171103. [CrossRef] [PubMed]

13. Bauswein, A.; Bastian, N.U.F.; Blaschke, D.B.; Chatziioannou, K.; Clark, J.A.; Fischer, T.; Oertel, M. Identifying a First-Order Phase Transition in Neutron-Star Mergers through Gravitational Waves. *Phys. Rev. Lett.* **2019**, *122*, 061102. [CrossRef] [PubMed]

14. Blacker, S.; Bastian, N.U.F.; Bauswein, A.; Blaschke, D.B.; Fischer, T.; Oertel, M.; Soultanis, T.; Typel, S. Constraining the onset density of the hadron-quark phase transition with gravitational-wave observations. *Phys. Rev. D* **2020**, *102*, 123023. [CrossRef]

15. Radice, D. General-relativistic Large-eddy Simulations of Binary Neutron Star Mergers. *Astrophys. J. Lett.* **2017**, *838*, L2. [CrossRef]

16. Breschi, M.; Bernuzzi, S.; Zappa, F.; Agathos, M.; Perego, A.; Radice, D.; Nagar, A. Kilohertz gravitational waves from binary neutron star remnants: Time-domain model and constraints on extreme matter. *Phys. Rev. D* **2019**, *100*, 104029. [CrossRef]

17. Bauswein, A.; Janka, H.T. Measuring Neutron-Star Properties via Gravitational Waves from Neutron-Star Mergers. *Phys. Rev. Lett.* **2012**, *108*, 011101. [CrossRef]

18. Hotokezaka, K.; Kiuchi, K.; Kyutoku, K.; Muranushi, T.; Sekiguchi, Y.i.; Shibata, M.; Taniguchi, K. Remnant massive neutron stars of binary neutron star mergers: Evolution process and gravitational waveform. *Phys. Rev. D* **2013**, *88*, 044026. [CrossRef]

19. Bernuzzi, S.; Nagar, A.; Balmelli, S.; Dietrich, T.; Ujevic, M. Quasiuniversal Properties of Neutron Star Mergers. *Phys. Rev. Lett.* **2014**, *112*, 201101. [CrossRef]

20. Rezzolla, L.; Takami, K. Gravitational-wave signal from binary neutron stars: A systematic analysis of the spectral properties. *Phys. Rev. D* **2016**, *93*, 124051. [CrossRef]

21. Zappa, F.; Bernuzzi, S.; Radice, D.; Perego, A.; Dietrich, T. Gravitational-Wave Luminosity of Binary Neutron Stars Mergers. *Phys. Rev. Lett.* **2018**, *120*, 111101. [CrossRef] [PubMed]

22. Most, E.R.; Papenfort, L.J.; Dexheimer, V.; Hanauske, M.; Schramm, S.; Stöcker, H.; Rezzolla, L. Signatures of Quark-Hadron Phase Transitions in General-Relativistic Neutron-Star Mergers. *Phys. Rev. Lett.* **2019**, *122*, 061101. [CrossRef] [PubMed]

23. Most, E.R.; Jens Papenfort, L.; Dexheimer, V.; Hanauske, M.; Stoecker, H.; Rezzolla, L. On the deconfinement phase transition in neutron-star mergers. *Eur. Phys. J. A* **2020**, *56*, 59. [CrossRef]

24. Liebling, S.L.; Palenzuela, C.; Lehner, L. Effects of high density phase transitions on neutron star dynamics. *Class. Quantum Gravity* **2021**, *38*, 115007. [CrossRef]

25. Prakash, A.; Radice, D.; Logoteta, D.; Perego, A.; Nedora, V.; Bombaci, I.; Kashyap, R.; Bernuzzi, S.; Endrizzi, A. Signatures of deconfined quark phases in binary neutron star mergers. *Phys. Rev. D* **2021**, *104*, 083029. [CrossRef]

26. Sebastian Blacker and Hristijan Kochankovski and Andreas Bauswein and Angels Ramos and Laura Tolos Thermal behavior as indicator for hyperons in binary neutron star merger remnants. *arXiv* **2023**, arXiv:2307.03710.

27. Mallick, R. Gravitational wave signatures of phase transition from hadronic to quark matter in isolated neutron stars and binaries. *Eur. Phys. J. Web Conf.* **2022**, *274*, 07002. [CrossRef]

28. Ren, J.; Zhang, C. Hybrid stars may have an inverted structure. *arXiv* **2022**, arXiv:2211.12043.

29. Tajima, H.; Tsutsui, S.; Doi, T.M.; Iida, K. Density-Induced Hadron–Quark Crossover via the Formation of Cooper Triples. *Symmetry* **2023**, *15*, 333. [CrossRef]

30. Kohsuke, S.; Toru, K.; Shun, F. Equation of State in Neutron Stars and Supernovae. In *Handbook of Nuclear Physics*; Springer: Singapore, 2023; pp. 1–51.

31. Sebastian, B.; Andreas, B.; Stefan, T. Exploring thermal effects of the hadron-quark matter transition in neutron star mergers. *arXiv* **2023**, arXiv:2304.01971.

32. Bauswein, A.; Janka, H.T.; Hebeler, K.; Schwenk, A. Equation-of-state dependence of the gravitational-wave signal from the ring-down phase of neutron-star mergers. *Phys. Rev. D* **2012**, *86*, 063001. [CrossRef]

33. Bauswein, A.; Blacker, S.; Vijayan, V.; Stergioulas, N.; Chatziioannou, K.; Clark, J.A.; Bastian, N.U.F.; Blaschke, D.B.; Cierniak, M.; Fischer, T. Equation of State Constraints from the Threshold Binary Mass for Prompt Collapse of Neutron Star Mergers. *Phys. Rev. Lett.* **2020**, *125*, 141103. [CrossRef] [PubMed]

34. Pisarski, R.D.; Skokov, V.V. Chiral matrix model of the semi-QGP in QCD. *Phys. Rev. D* **2016**, *94*, 034015. [CrossRef]

35. Steinheimer, J.; Schramm, S. The problem of repulsive quark interactions - Lattice versus mean field models. *Phys. Lett. B* **2011**, *696*, 257–261. [CrossRef]

36. Hatsuda, T.; Tachibana, M.; Yamamoto, N.; Baym, G. New Critical Point Induced By the Axial Anomaly in Dense QCD. *Phys. Rev. Lett.* **2006**, *97*, 122001. [CrossRef]

37. Aoki, Y.; Endrődi, G.; Fodor, Z.; Katz, S.D.; Szabó, K.K. The order of the quantum chromodynamics transition predicted by the standard model of particle physics. *Nature* **2006**, *443*, 675–678. [CrossRef]

38. Baym, G.; Hatsuda, T.; Kojo, T.; Powell, P.D.; Song, Y.; Takatsuka, T. From hadrons to quarks in neutron stars: A review. *Rep. Prog. Phys.* **2018**, *81*, 056902. [CrossRef]

39. Baym, G.; Furusawa, S.; Hatsuda, T.; Kojo, T.; Togashi, H. New Neutron Star Equation of State with Quark-Hadron Crossover. *Astrophys. J.* **2019**, *885*, 42. [CrossRef]

40. Kojo, T.; Baym, G.; Hatsuda, T. Implications of NICER for Neutron Star Matter: The QHC21 Equation of State. *Astrophys. J.* **2022**, *934*, 46. [CrossRef]

41. Sathyaprakash, B.; Abernathy, M.; Acernese, F.; Ajith, P.; Allen, B.; Amaro-Seoane, P.; Andersson, N.; Aoudia, S.; Arun, K.; Astone, P.; et al. Scientific objectives of Einstein Telescope. *Class. Quantum Gravity* **2012**, *29*, 124013. [CrossRef]

42. Abbott, B.P.; Abbott, R.; Abbott, T.D.; Abernathy, M.R.; Ackley, K.; Adams, C.; Addesso, P.; Adhikari, R.X.; Adya, V.B.; Affeldt, C.; et al. Exploring the sensitivity of next generation gravitational wave detectors. *Class. Quantum Gravity* **2017**, *34*, 044001. [CrossRef]
43. Hatsuda, T.; Kunihiro, T. QCD phenomenology based on a chiral effective Lagrangian. *Phys. Rep.* **1994**, *247*, 221–367. [CrossRef]
44. Alford, M.G.; Schmitt, A.; Rajagopal, K.; Schäfer, T. Color superconductivity in dense quark matter. *Rev. Mod. Phys.* **2008**, *80*, 1455–1515. [CrossRef]
45. Chabanat, E.; Bonche, P.; Haensel, P.; Meyer, J.; Schaeffer, R. A Skyrme parametrization from subnuclear to neutron star densitiesPart II. Nuclei far from stabilities. *Nucl. Phys. A* **1998**, *635*, 231–256. [CrossRef]
46. Glendenning, N.K. Neutron stars are giant hypernuclei? *Astrophys. J.* **1985**, *293*, 470–493. [CrossRef]
47. Nambu, Y.; Jona-Lasinio, G. Dynamical Model of Elementary Particles Based on an Analogy with Superconductivity. I. *Phys. Rev.* **1961**, *122*, 345–358. [CrossRef]
48. Nambu, Y.; Jona-Lasinio, G. Dynamical Model of Elementary Particles Based on an Analogy with Superconductivity. II. *Phys. Rev.* **1961**, *124*, 246–254. [CrossRef]
49. Buballa, M. NJL-model analysis of dense quark matter [review article]. *Phys. Rep.* **2005**, *407*, 205–376. [CrossRef]
50. Etienne, Z.; Brandt, S.R.; Diener, P.; Gabella, W.E.; Gracia-Linares, M.; Haas, R.; Kedia, A.; Alcubierre, M.; Alic, D.; Allen, G.; et al. The Einstein Toolkit. *Zenodo (The "Lorentz" release, ET_2021_05)* **2021**. [CrossRef]
51. Nakamura, T.; Oohara, K.; Kojima, Y. General Relativistic Collapse to Black Holes and Gravitational Waves from Black Holes. *Prog. Theor. Phys. Suppl.* **1987**, *90*, 1–218. [CrossRef]
52. Shibata, M.; Nakamura, T. Evolution of three-dimensional gravitational waves: Harmonic slicing case. *Phys. Rev. D* **1995**, *52*, 5428–5444. [CrossRef] [PubMed]
53. Baumgarte, T.W.; Shapiro, S.L. Numerical integration of Einstein's field equations. *Phys. Rev. D* **1998**, *59*, 024007. [CrossRef]
54. Alcubierre, M.; Brügmann, B.; Dramlitsch, T.; Font, J.A.; Papadopoulos, P.; Seidel, E.; Stergioulas, N.; Takahashi, R. Towards a stable numerical evolution of strongly gravitating systems in general relativity: The conformal treatments. *Phys. Rev. D* **2000**, *62*, 044034. [CrossRef]
55. Alcubierre, M.; Brügmann, B.; Diener, P.; Koppitz, M.; Pollney, D.; Seidel, E.; Takahashi, R. Gauge conditions for long-term numerical black hole evolutions without excision. *Phys. Rev. D* **2003**, *67*, 084023. [CrossRef]
56. Baiotti, L.; Hawke, I.; Montero, P.J.; Löffler, F.; Rezzolla, L.; Stergioulas, N.; Font, J.A.; Seidel, E. Three-dimensional relativistic simulations of rotating neutron-star collapse to a Kerr black hole. *Phys. Rev. D* **2005**, *71*, 024035. [CrossRef]
57. Hawke, I.; Löffler, F.; Nerozzi, A. Excision methods for high resolution shock capturing schemes applied to general relativistic hydrodynamics. *Phys. Rev. D* **2005**, *71*, 104006. [CrossRef]
58. Mösta, P.; Mundim, B.C.; Faber, J.A.; Haas, R.; Noble, S.C.; Bode, T.; Löffler, F.; Ott, C.D.; Reisswig, C.; Schnetter, E. GRHydro: A new open-source general-relativistic magnetohydrodynamics code for the Einstein toolkit. *Class. Quantum Gravity* **2014**, *31*, 015005. [CrossRef]
59. Banyuls, F.; Font, J.A.; Ibáñez, J.M.; Martí, J.M.; Miralles, J.A. Numerical {3 + 1} General Relativistic Hydrodynamics: A Local Characteristic Approach. *Astrophys. J.* **1997**, *476*, 221–231. [CrossRef]
60. Font, J.A. Numerical Hydrodynamics and Magnetohydrodynamics in General Relativity. *Living Rev. Relativ.* **2008**, *11*, 7. [CrossRef]
61. Gourgoulhon, E.; Grandclément, P.; Taniguchi, K.; Marck, J.A.; Bonazzola, S. Quasiequilibrium sequences of synchronized and irrotational binary neutron stars in general relativity: Method and tests. *Phys. Rev. D* **2001**, *63*, 064029. [CrossRef]
62. Grandclément, P.; Novak, J. Spectral Methods for Numerical Relativity. *Living Rev. Relativ.* **2009**, *12*, 1. [CrossRef] [PubMed]
63. Schnetter, E.; Hawley, S.H.; Hawke, I. Evolutions in 3D numerical relativity using fixed mesh refinement. *Class. Quantum Gravity* **2004**, *21*, 1465–1488. [CrossRef]
64. Schnetter, E.; Diener, P.; Dorband, E.N.; Tiglio, M. A multi-block infrastructure for three-dimensional time-dependent numerical relativity. *Class. Quantum Gravity* **2006**, *23*, S553–S578. [CrossRef]
65. De Pietri, R.; Feo, A.; Maione, F.; Löffler, F. Modeling equal and unequal mass binary neutron star mergers using public codes. *Phys. Rev. D* **2016**, *93*, 064047. [CrossRef]
66. Takami, K.; Rezzolla, L.; Baiotti, L. Spectral properties of the post-merger gravitational-wave signal from binary neutron stars. *Phys. Rev. D* **2015**, *91*, 064001. [CrossRef]

universe

Review

Solving the Mystery of Fast Radio Bursts: A Detective's Approach

Bing Zhang [1,2]

1   Nevada Center for Astrophysics, University of Nevada, Las Vegas, NV 89154-4002, USA;
     bing.zhang@unlv.edu
2   Department of Physics and Astronomy, University of Nevada, Las Vegas, NV 89154-4002, USA

Abstract: Fast radio bursts (FRBs) are still a mystery in contemporary astrophysics. Unlike many other astronomical objects whose basic physical mechanism is already identified and the research on which focuses mainly on refining details, FRBs are still largely unknown regarding their source(s) and radiation mechanism(s). To make progress in the field, a "top-down" or "detective's approach" is desirable. I will summarize how some key observational facts have narrowed down the options to interpret FRBs and show that at least some FRBs are produced from the magnetospheres of highly magnetized neutron stars (or magnetars). I will also argue that the current data seem to favor a type of coherent inverse Compton scattering process by relativistic particle bunches off a low-frequency wave propagating in the magnetosphere. This brief contribution is a shorter version of an extended review to be published in *Reviews of Modern Physics*, and it was written as a tribute to the 80th anniversary of Remo Ruffini.

Keywords: fast radio bursts; magnetars; coherent radio emission

1. Prologue

The 80th birthday of Prof. Remo Ruffini was on 17 May 2022. Because of the COVID-19 pandemic, I was not able to attend the dedicated celebration conference in person but was invited to deliver a remote talk. Incidentally, I had the pleasure of celebrating his 77th birthday three years earlier at the 2019 Nanjing GRB conference. I have known Remo for many years, meeting with him at numerous conferences in high-energy astrophysics. Our research overlaps in the field of gamma-ray bursts (GRBs). Even though we often interpret GRB phenomenology differently, I have enjoyed many conversations with him about the physics of GRBs and other subjects.

Our different views on GRBs stem from the different approaches we have taken to tackle the GRB problem. As a distinguished relativistist, Remo often adopts a doctrinal or "bottom-up" approach by setting up a theoretical framework to begin with and matching observations with the theories. The examples include his theory of electromagnetic black holes and the fireshell model for GRBs (e.g., [1–3]) to interpret GRB prompt emission and afterglow, and his progenitor models involving a list of binary systems (e.g., [4]). The approach I and many others take is the opposite. We start with the observational data and ask ourselves what the data really tell us. By ruling out various possibilities (including some of our own ideas that were proposed before the relevant data became available), we finally narrowed down the most probable interpretations of the phenomenon. Such a "top-down" approach is analogous to the approach of a detective who tries to unveil a crime scene. My understanding of the GRB phenomenology has been summarized in the book titled "The Physics of Gamma-Ray Bursts" [5].

In research fields such as GRBs and fast radio bursts (FRBs), whose data are quite sparse in the early stage of development, I believe that the detective's approach is more fruitful. Our initial bets usually turn out incorrect, and new surprising discoveries keep flooding in, forcing continuous revisions of the theoretical framework.

The talk I remotely presented at Ruffini's 80th birthday meeting was titled "The Physics of Fast Radio Bursts". The content of the talk has been explained in detail in my

Citation: Zhang, B. Solving the Mystery of Fast Radio Bursts: A Detective's Approach. *Universe* **2023**, *9*, 375. https://doi.org/10.3390/universe9080375

Academic Editors: Luciano Nicastro, Remo Ruffini, Jorge Armando Rueda Hernández, Narek Sahakyan and Gregory Vereshchagin

Received: 19 May 2023
Revised: 10 August 2023
Accepted: 14 August 2023
Published: 18 August 2023

long review article in press at the *Reviews of Modern Physics* [6]. This short contribution is a highlight of that review, with a focus on how the detective's approach bears fruit in this rapidly developing field.

2. The Source(s) of FRBs

FRBs [7,8] were first reported in 2007 as highly dispersed, millisecond-duration bursts detected in the radio band (from $\sim$110 MHz to $\sim$8 GHz), see also an earlier controversial case [9]. According to [10] and the FRB theory catalog (https://frbtheorycat.org/index.php/Main_Page, accessed on 13 June 2019), there are more than 50 models proposed in the literature. Most of these models have been critically commented on in [6], with most of them already disfavored by the data. In the following, I will list the key observations that have greatly narrowed down the possible source models to interpret FRBs.

The key observational clues that are related to the source(s) of FRBs include the following:

- The smoking gun: An MJy radio burst (FRB 200428) was observed from the Galactic magnetar SGR J1935+2154, which was temporally associated with a moderately bright X-ray burst [11–16]. The radio burst, if observed from nearby galaxies, would appear as a low-luminosity FRB. This suggests that at least magnetars can make FRBs, and at least some FRBs are made by magnetars. The majority of X-ray bursts emitted from this source were, however, NOT associated with FRBs [17], suggesting that special conditions are needed for a magnetar to make FRB-like events. Later, a radio pulsar phase was observed from the magnetar. The pulses are found to be confined in a narrow phase window. FRB bursts, on the other hand, appear in random rotation phases, suggesting that the bursts and pulses likely originate from different locations and probably have somewhat different mechanisms (albeit sharing similar physics) [18].
- Cosmological FRBs are observed to have two apparent types: repeaters and non-repeaters. There have been intense discussions regarding whether all FRBs repeat (e.g., [19–22]). Some observations show that repeating bursts have some special features (e.g., broader pulses and narrower spectra [23]). However, as the observing time increases, some previously named non-repeaters turn into repeaters [24]. The separation between the two populations becomes more blurred.
- There are several very active repeaters that, when monitored closely with the Five-hundred-meter Aperture Spherical radio Telescope (FAST) in China, have a burst rate exceeding 100 per hour [25–29].
- Two active repeaters, FRB 20121102A [30] and FRB 20190520B [31], are located inside a persistent radio source (PRS). Other repeaters, on the other hand, do not have detectable PRSs. The two PRS repeaters also have large Faraday rotation measure (RM) values, suggesting a possible dense and highly magnetized environment. The RM values of some active repeaters also undergo significant long-term [32] and short-term [26] variations, sometimes with significant sign reversals [33,34]. All these are, however, not necessary conditions to produce an active repeater. FRB 20220912A has a negligibly small and non-varying RM yet actively emits many bursts with the total burst energy budget comparable to other active repeaters [29].
- The host galaxies of FRBs (both repeaters and apparent non-repeaters) seem to be mostly Milky-Way-like massive galaxies, unlike the star-forming dwarf host galaxies of long GRBs and superluminous supernovae [35–42]. The positions of FRBs within the host galaxies also typically have large offsets from the star-forming regions [26,38,39]. The global properties are more analogous to Type II supernovae, Type Ia supernovae, and even short GRBs [43]. The DM distribution of the FRBs from the CHIME first catalog seems to require a delayed channel from star formation, at least for some FRBs [44–46], even though the star formation model is consistent with the data if some nearby (low DM) samples are removed [47].
- The existence of FRBs with a delayed channel is solidified by the discovery of FRB 20200120E in a globular cluster of a nearby spiral galaxy M81 at a distance of

3.6 Mpc [48,49]. The bursts from the source have lower luminosities than typical cosmological FRBs, suggesting that there could be many more such sources from far away galaxies that have evaded detection [50–52].

- Most repeaters do not have a detectable apparent periodicity. There are two special cases detected by the CHIME/FRB collaboration: (1) FRB 20180916B has an apparent $\sim$16 day periodicity [53] with frequency-dependent active phases [54,55]. Close monitoring of other active repeaters with the FAST telescope does not show significant periodic signals (the tentative $\sim$157-day period for FRB 20121102A [56] does not increase significance with time P. Wang et al. 2023, in prep). (2) FRB 20191221A, was identified to show a 216.8(1) ms periodicity with a significance of 6.5σ [57], but since it has a roughly 3 s-long duration which is much longer than the typical millisecond duration of other FRBs, this FRB is likely a special case and may have a different origin from the majority of FRBs.

So what do these clues tell us about the FRB sources? Here is a list of statements one may make after performing a detective's analysis:

- Magnetars can make FRBs;
- At least some FRBs are produced by moderate-age magnetars such as SGR J1935+2154. Because the source of FRB 200428 (SGR J1935+2154) continues to emit more X-ray bursts later, the magnetar FRB sources must be FRB repeaters;
- Since we have not detected an active FRB repeater in the Milky Way galaxy, the active repeaters at cosmological distances may require a different interpretation. Additionally, since none of the Galactic magnetars are known to be located in globular clusters, the source of FRB 20200120E must be somewhat different from the source of FRB 200428.
- For the above reasons, overall, there may exist at least three types or sub-types of repeating FRB sources: SGR J1935+2154-like magnetars, active repeater sources, and globular cluster sources.
- If magnetars are the common engine for all repeating FRBs, then there might be at least three sub-categories of magnetars: the common ones such as SGR J1935+2154 that make FRB 200428-like bursts with a low repetition rate, the special magnetars (presumably younger) that power active repeaters (with the caveat that some of these active repeaters are also located in regions offset from the main galaxy light), and more special magnetars newly born in globular clusters.
- An alternative, more speculative but exciting possibility is that at least some cosmological FRBs do not originate from self-bursting magnetars. Interacting neutron stars or even black hole systems are other well-motivated possibilities and cannot be easily ruled out with the current data.
- The immediate environment of active repeaters can be either highly magnetized and dynamically evolving or the opposite. In some cases, the environment is consistent with a dense magnetized nebula or a magnetized companion in a binary system, but in some other cases, there is no evidence of any of such a complicated environment. This suggests that the emission sources of FRBs should not rely heavily on the environment to produce the bursts.
- There are no definite clues yet that some apparent non-repeaters must originate from catastrophic events. A possible case of intrinsic one-off FRB from a plausible association between an FRB and a gravitational wave event has been suggested [58], but the case is controversial: Ref. [59] suggested that the association is physically impossible if the host galaxy is the one suggested in [60]. More observations are needed to see whether another distinct class of FRBs can be established.

3. The Radiation Mechanism(s) of FRBs

Because of their high fluxes, short durations, and large distances, FRBs have the most extreme high brightness temperatures in the universe, reaching $T_b \sim 10^{36}$ K and higher. This raises great challenges in identifying their coherent radiation mechanism(s). In [6], I have reviewed in depth various FRB radiation models proposed in the literature and discussed the pros and cons of these models in confrontation with the data. Here I just summarize the main observations and the constraints posed by them.

The key observational clues that are related to the radiation mechanism(s) of FRBs include the following:

- The typical duration of FRBs is milliseconds, but some FRBs have rapid variability as short as 60 nanoseconds [50].
- Most observed FRBs are highly polarized. Most have nearly 100% linear polarization [32,61–63]. A small fraction of bursts from a growing population of sources have measurable circular polarization, some even up to 70% [26,64–68].
- Most bursts with high linear polarization degrees have polarization angle (PA) nearly constant during the bursts [32,69], but cases with significant PA variations have been observed in bursts from some sources [61,62,69].
- The isotropic burst energy emitted in the radio band during the timescale of an active episode of repeater sources (ranging from a few days to 1–2 months) is of the order of a few $\sim 10^{43}$ erg [25–29].
- Some FRBs, especially non-repeaters, show wide spectra with emissions covering the entire bandpass of the telescopes. For repeaters, on the other hand, the spectra are typically narrow [23]. Case studies of many bursts from FRB 20201124A [27], and FRB 20220912A [29] suggest that some bursts even have $\delta\nu/\nu_0$ as small as <0.3.
- Some bursts, especially those of repeaters, show an interesting frequency down-drifting feature (also called the "sad-trombone" effect), with higher frequency emission arriving earlier and lower frequency emission arriving later [23,27,70].

FRB emission models within the magnetar framework may be generally grouped into two categories. The first category borrows insight from modeling radio pulsars and invokes pulsar magnetospheres (either inside the magnetosphere or slightly outside the magnetosphere in the current sheet region beyond the light cylinder), which may be called "closer-in" or "pulsar-like" models. The second category invokes highly magnetized relativistic shocks far from the engine. The physical processes in such a scenario share some aspects with GRBs. Such models may be termed "farther-out" or "GRB-like" models. These models have been discussed in great detail in [6] with many original papers cited. In the following section, I summarize some key constraints on these models based on the observational facts listed above.

- Polarization angle (PA) swing is a key observational feature of radio pulsars. As the line of sight sweeps across different field lines when the neutron star rotates, different PAs are observed. The characteristic signature is an "S" shape or its inverse, which is consistent with the dipolar geometry of magnetic fields conjectured for pulsars. The variation becomes smaller if the line of sight tangentially cuts the emission cone or the emission height is large. Conversely, the synchrotron maser model invoking a magnetized relativistic shock demands parallel magnetic field lines to achieve coherence and, therefore, only predicts non-varying PAs across a burst. Such non-varying PAs are indeed observed in most bursts [32,69], but the detection of varying PAs [61,63] from both repeating and non-repeating FRBs rules out the shock model at least for some FRBs. Since the magnetospheric models can account for both varying and non-varying PAs, diverse PA variations offer strong support to the magnetospheric origin of FRBs.
- Circular polarization [26,68,69] can be produced either from intrinsic radiation mechanisms or propagation effects [71]. The detection of significant circular polarization from a large fraction of bursts from the clean-environment active repeater FRB 20220912A [29] suggests that circular polarization is very likely unrelated to the propagation effect in the external medium. Since synchrotron maser emission cannot produce bright bursts with significant circular polarization [71] whereas magnetospheric models can do so [71–73] via coherent curvature [74,75] or inverse Compton scattering [76] or through propagation effects within the magnetosphere [71], the

circular polarization data therefore also disfavor the GRB-like models and favor the pulsar-like models.

- The 60-ns variability timescale [50] poses a significant challenge to the GRB-like shock model [51]. The duration of the burst w defines an FRB emission radius $R_{FRB} \sim cw\Gamma^2$. A variability timescale $\delta t \ll w$ would have to be attributed to small patches in a relativistic jet, which introduces a very low efficiency for emission [77]. This argument was well known in the GRB field to argue against the external shock model for GRB prompt emission.

- The narrow spectrum $\delta\nu/\nu_0 < 0.3$ observed in some repeaters [27,29] poses a generic constraint on the GRB-like model, and even challenges most of the pulsar-like models ([78], Y. Qu, P. Kumar and B. Zhang, 2023, in preparation).

- The relativistic shock model predicts a small radio emission efficiency of the order of 10^{-4} [79]. Pulsar-like models are more flexible since the pulsar radio emission efficiency can range from 10^{-7} to ~ 1 [80]. The Galactic FRB 200428 has an efficiency of the order of 10^{-4}, which can be accounted for in both models. However, if extragalactic FRBs have the similar efficiency, even if there is a global beaming factor $f_B \sim 0.1$, the observed $\sim 10^{43}$ erg isotropic radio burst emission energy measured during active episodes for a few repeaters would suggest a total energy of a few times 10^{46} erg within an active episode, which is already a significant fraction of the dipolar magnetic energy of a magnetar [25–29]. This suggests that if extragalactic FRBs are powered by magnetars, the GRB-like models already suffer from the energetics problem. The pulsar-like models are still allowed if they can make FRBs much more efficiently than the 10^{-4} efficiency.

- The sad-trombone effect can be naturally interpreted within the pulsar-like model using the "radius-to-frequency-mapping" effect widely discussed in pulsar models [81,82]. This effect can also be accounted for within the shock model, even though some special conditions are required [83,84].

In summary, many independent pieces of evidence point toward a consistent picture in favor of a magnetospheric origin of at least some, and probably most, FRBs if magnetars are the common source engine of FRBs.

4. Magnetospheric Coherent Inverse Compton Scattering as an Attractive Mechanism to Power FRB Emission

Within the magnetospheric models, based on the observational clues collected so far, I personally favor a mechanism invoking the upscattering of a certain type of low-frequency waves by relativistic particle bunches within the magnetosphere of a magnetar. The simplest scenario is to invoke a low-frequency electromagnetic wave, which might be excited by oscillations of near-surface charges induced by crust cracking [76]. The advantages of this model in interpreting the observations include the following:

- It provides an alternative to the traditional model invoking coherent curvature radiation by bunches. It inherits the merit of that model that invokes the magnetar magnetosphere as the emission site but can overcome some difficulties encountered by the curvature radiation model.

- One difficulty of the coherent curvature radiation model is the plasma suppression effect. Because of the huge brightness temperatures of FRBs, the required bunches need to have a very large plasma density (or a high multiplicity with respect to the Goldreich–Julian density), which exacerbates the plasma suppression effect that has been discussed within the context of pulsar emission [85]. The emission power of the inverse Compton scattering (ICS) process for individual particles is a few orders of magnitude larger than that of curvature radiation, which suggests that a dense plasma is not needed for such a mechanism to power bright FRB emission [76]. The plasma suppression effect is no longer relevant. Even for the curvature radiation model, the plasma suppression effect may become irrelevant when one considers a parallel electric field in the emission region, which will separate the opposite charges in the plasma [86]. Such an $E_\parallel$ is needed in FRB models invoking coherent radiation by bunches (for both curvature radiation and ICS) in order to overcome efficient cooling of the bunches to maintain a large enough power to produce an FRB [74,76] and mandate a low-twist magnetar model [87].

- Another major difficulty for coherent curvature radiation by bunches is how to produce and maintain bunches with the right size that corresponds to the observed wavelength. Two stream instabilities have been widely invoked to generate bunches, but the characteristic frequencies for those instabilities involve the plasma frequency, which is not obvious how it is related to the observed FRB frequency. The ICS process provides a way to naturally bunch the particles. In the rest-frame of a relativistically moving particle, the electric field of the incident wave (which is Dopper boosted from the low-frequency wave) provides an oscillating force that naturally bunches the particles in the scale of the wavelength in the comoving frame. The emitted electromagnetic wave (due to Thomson scattering in the comoving frame) carries the same frequency, which is Doppler boosted in the observer frame to the FRB frequency. As a result, the emitting particles are naturally bunched within the wavelength of the emitted waves. This provides a very natural bunching mechanism not shared by the curvature radiation mechanism.
- The maintenance of the bunches within the FRB models (both for curvature radiation and ICS) is attributed to the $E_\parallel$ in the emission region, which is demanded by the energetics argument [74,76]. Such an $E_\parallel$ allows the bunches to emit in the radiation-reaction-limited regime. In such a regime, the particle energy distribution would maintain a narrow distribution due to the "thermostat" effect, i.e., the particles with a larger energy than the critical Lorentz factor defined by the radiation-reaction-limit condition would undergo stronger radiative cooling to lose energy, whereas the particles with lower energy than the critical Lorentz factor would undergo further acceleration via $E_\parallel$. As a result, the particle distribution within the bunch would maintain a value around the critical Lorentz factor so that the bunch is not easily dispersed.
- Curvature radiation is intrinsically a wide-band spectrum characterized by the modified Bessel's function for a single electron. The ICS process, on the other hand, could produce a much narrower spectrum for a single electron, given that the low-frequency wave itself has a narrow spectrum (which is possible since its frequency may correspond to a characteristic mode of neutron star crustal oscillations). As a result, the ICS process has a better prospect of producing narrow spectra, as observed than the curvature radiation model.

5. Epilogue

The history of GRB studies has shown the power of the top-down approach. With the combination of observational data and theoretical modeling, a standard physical framework has been established for GRBs, with the final major piece of the GRB paradigm collected in 2017 (the gravitational wave - short GRB association) 50 years after the discovery of the first GRB. The field of FRB research shares a similar history as the GRB field but with an expedited pace. At the time of writing, we are still in the process of putting together a coherent physical picture for FRBs. The bullet points summarized in the previous sections are my best guess as a detective at the stage of writing. It is almost certain that some of these statements will turn out incorrect when more observations bring further clues. This contribution, in any case, serves as an intermediate record that may be reviewed and entertained later.

Funding: The author acknowledges Nevada Center for Astrophysics and NASA 80NSSC23M0104 for support.

Data Availability Statement: Not applicable.

Conflicts of Interest: The author declares no conflict of interest.

References

1. Ruffini, R.; Bernardini, M.G.; Bianco, C.L.; Chardonnet, P.; Fraschetti, F.; Gurzadyan, V.; Vitagliano, L.; Xue, S.S. The Blackholic energy: Long and short Gamma-Ray Bursts (New perspectives in physics and astrophysics from the theoretical understanding of Gamma-Ray Bursts, II). In *American Institute of Physics Conference Series, Proceedings of the XIth Brazilian School of Cosmology and Gravitation, Rio de Janeiro, Brazil, 26 July–4 August 2004*; Novello, M., Perez Bergliaffa, S.E., Eds.; American Institute of Physics: College Park, MD, USA, 2005; Volume 782, pp. 42–127. [CrossRef]
2. Ruffini, R.; Bernardini, M.G.; Bianco, C.L.; Caito, L.; Chardonnet, P.; Dainotti, M.G.; Fraschetti, F.; Guida, R.; Rotondo, M.; Vereshchagin, G.; et al. The Blackholic energy and the canonical Gamma-Ray Burst. In *American Institute of Physics Conference Series, Proceedings of the XIIth Brazilian School of Cosmololy and Gravitation, Rio de Janeiro, Brazil, 20 July–2 August 2008*; Novello, M., Perez Bergliaffa, S.E., Eds.; American Institute of Physics: College Park, MD, USA, 2007; Volume 910, pp. 55–217. [CrossRef]
3. Ruffini, R.; Vereshchagin, G.; Xue, S.S. Electron-positron pairs in physics and astrophysics: From heavy nuclei to black holes. *Phys. Rep.* **2010**, *487*, 1–140. [CrossRef]
4. Ruffini, R.; Rueda, J.A.; Muccino, M.; Aimuratov, Y.; Becerra, L.M.; Bianco, C.L.; Kovacevic, M.; Moradi, R.; Oliveira, F.G.; Pisani, G.B.; et al. On the Classification of GRBs and Their Occurrence Rates. *Astrophys. J.* **2016**, *832*, 136. [CrossRef]
5. Zhang, B. *The Physics of Gamma-Ray Bursts*; Cambridge University Press: Cambridge, UK, 2018. [CrossRef]
6. Zhang, B. The Physics of Fast Radio Bursts. *arXiv* **2022**, arXiv:2212.03972. [CrossRef]
7. Lorimer, D.R.; Bailes, M.; McLaughlin, M.A.; Narkevic, D.J.; Crawford, F. A Bright Millisecond Radio Burst of Extragalactic Origin. *Science* **2007**, *318*, 777. [CrossRef]
8. Thornton, D.; Stappers, B.; Bailes, M.; Barsdell, B.; Bates, S.; Bhat, N.D.R.; Burgay, M.; Burke-Spolaor, S.; Champion, D.J.; Coster, P.; et al. A Population of Fast Radio Bursts at Cosmological Distances. *Science* **2013**, *341*, 53–56. [CrossRef]
9. Linscott, I.R.; Erkes, J.W. Discovery of millisecond radio bursts from M 87. *Astrophys. J.* **1980**, *236*, L109–L113. [CrossRef]
10. Platts, E.; Weltman, A.; Walters, A.; Tendulkar, S.P.; Gordin, J.E.B.; Kandhai, S. A living theory catalogue for fast radio bursts. *Phys. Rep.* **2019**, *821*, 1–27.
11. Andersen, B.C. et al. [CHIME/FRB Collaboration] A bright millisecond-duration radio burst from a Galactic magnetar. *Nature* **2020**, *587*, 54–58.
12. Bochenek, C.D.; Ravi, V.; Belov, K.V.; Hallinan, G.; Kocz, J.; Kulkarni, S.R.; McKenna, D.L. A fast radio burst associated with a Galactic magnetar. *Nature* **2020**, *587*, 59–62.
13. Li, C.K.; Lin, L.; Xiong, S.L.; Ge, M.Y.; Li, X.B.; Li, T.P.; Lu, F.J.; Zhang, S.N.; Tuo, Y.L.; Nang, Y.; et al. HXMT identification of a non-thermal X-ray burst from SGR J1935+2154 and with FRB 200428. *Nat. Astron.* **2021**, *5*, 378.
14. Mereghetti, S.; Savchenko, V.; Ferrigno, C.; Götz, D.; Rigoselli, M.; Tiengo, A.; Bazzano, A.; Bozzo, E.; Coleiro, A.; Courvoisier, T.J.L.; et al. INTEGRAL Discovery of a Burst with Associated Radio Emission from the Magnetar SGR 1935+2154. *Astrophys. J.* **2020**, *898*, L29.
15. Ridnaia, A.; Svinkin, D.; Frederiks, D.; Bykov, A.; Popov, S.; Aptekar, R.; Golenetskii, S.; Lysenko, A.; Tsvetkova, A.; Ulanov, M.; et al. A peculiar hard X-ray counterpart of a Galactic fast radio burst. *Nat. Astron.* **2021**, *5*, 372–377.
16. Tavani, M.; Casentini, C.; Ursi, A.; Verrecchia, F.; Addis, A.; Antonelli, L.A.; Argan, A.; Barbiellini, G.; Baroncelli, L.; Bernardi, G.; et al. An X-ray burst from a magnetar enlightening the mechanism of fast radio bursts. *Nat. Astron.* **2021**, *5*, 401–407.
17. Lin, L.; Zhang, C.F.; Wang, P.; Gao, H.; Guan, X.; Han, J.L.; Jiang, J.C.; Jiang, P.; Lee, K.J.; Li, D.; et al. No pulsed radio emission during a bursting phase of a Galactic magnetar. *Nature* **2020**, *587*, 63–65.
18. Zhu, W.; Xu, H.; Zhou, D.; Lin, L.; Wang, B.; Wang, P.; Zhang, C.; Niu, J.; Chen, Y.; Li, C.; et al. A radio pulsar phase from SGR J1935+2154 provides clues to the magnetar FRB mechanism. *Sci. Adv.* **2023**, *9*, eadf6198.
19. Palaniswamy, D.; Li, Y.; Zhang, B. Are There Multiple Populations of Fast Radio Bursts? *Astrophys. J.* **2018**, *854*, L12.
20. Caleb, M.; Stappers, B.W.; Rajwade, K.; Flynn, C. Are all fast radio bursts repeating sources? *Mon. Not. R. Astron. Soc.* **2019**, *484*, 5500–5508.
21. Ravi, V. The prevalence of repeating fast radio bursts. *Nat. Astron.* **2019**, *3*, 928–931.
22. Ai, S.; Gao, H.; Zhang, B. On the True Fractions of Repeating and Nonrepeating Fast Radio Burst Sources. *Astrophys. J.* **2021**, *906*, L5.
23. Andersen, B.C. et al. [CHIME/FRB Collaboration] CHIME/FRB Discovery of Eight New Repeating Fast Radio Burst Sources. *Astrophys. J.* **2019**, *885*, L24.
24. Andersen, B.C. et al. [CHIME/FRB Collaboration] CHIME/FRB Discovery of 25 Repeating Fast Radio Burst Sources. *arXiv* **2023**, arXiv:2301.08762. [CrossRef]
25. Li, D.; Wang, P.; Zhu, W.W.; Zhang, B.; Zhang, X.X.; Duan, R.; Zhang, Y.K.; Feng, Y.; Tang, N.Y.; Chatterjee, S.; et al. A bimodal burst energy distribution of a repeating fast radio burst source. *Nature* **2021**, *598*, 267–271.
26. Xu, H.; Niu, J.R.; Chen, P.; Lee, K.J.; Zhu, W.W.; Dong, S.; Zhang, B.; Jiang, J.C.; Wang, B.J.; Xu, J.W.; et al. A fast radio burst source at a complex magnetized site in a barred galaxy. *Nature* **2022**, *609*, 685–688.
27. Zhou, D.J.; Han, J.L.; Zhang, B.; Lee, K.J.; Zhu, W.W.; Li, D.; Jing, W.C.; Wang, W.Y.; Zhang, Y.K.; Jiang, J.C.; et al. FAST observations of an extremely active episode of FRB 20201124A: I. Burst morphology. *arXiv* **2022**, arXiv:2210.03607.
28. Zhang, Y.K.; Wang, P.; Feng, Y.; Zhang, B.; Li, D.; Tsai, C.W.; Niu, C.H.; Luo, R.; Yao, J.M.; Zhu, W.W.; et al. FAST observations of an extremely active episode of FRB 20201124A: II. Energy Distribution. *arXiv* **2022**, arXiv:2210.03645.

29. Zhang, Y.K.; Li, D.; Zhang, B.; Cao, S.; Feng, Y.; Wang, W.Y.; Qu, Y.H.; Niu, J.R.; Zhu, W.W.; Han, J.L.; et al. FAST Observations of FRB 20220912A: Burst Properties and Polarization Characteristics. *arXiv* **2023**, arXiv:2304.14665. [CrossRef]
30. Chatterjee, S.; Law, C.J.; Wharton, R.S.; Burke-Spolaor, S.; Hessels, J.W.T.; Bower, G.C.; Cordes, J.M.; Tendulkar, S.P.; Bassa, C.G.; Demorest, P.; et al. A direct localization of a fast radio burst and its host. *Nature* **2017**, *541*, 58–61.
31. Niu, C.H.; Aggarwal, K.; Li, D.; Zhang, X.; Chatterjee, S.; Tsai, C.W.; Yu, W.; Law, C.J.; Burke-Spolaor, S.; Cordes, J.M.; et al. A repeating fast radio burst in a dense environment with a compact persistent radio source. *arXiv* **2021**, arXiv:2110.07418.
32. Michilli, D.; Seymour, A.; Hessels, J.W.T.; Spitler, L.G.; Gajjar, V.; Archibald, A.M.; Bower, G.C.; Chatterjee, S.; Cordes, J.M.; Gourdji, K.; et al. An extreme magneto-ionic environment associated with the fast radio burst source FRB 121102. *Nature* **2018**, *553*, 182–185.
33. Anna-Thomas, R.; Connor, L.; Burke-Spolaor, S.; Beniamini, P.; Aggarwal, K.; Law, C.J.; Lynch, R.S.; Li, D.; Feng, Y.; Ocker, S.K.; et al. A Highly Variable Magnetized Environment in a Fast Radio Burst Source. *arXiv* **2022**, arXiv:2202.11112.
34. Dai, S.; Feng, Y.; Yang, Y.P.; Zhang, Y.K.; Li, D.; Niu, C.H.; Wang, P.; Xue, M.Y.; Zhang, B.; Burke-Spolaor, S.; et al. Magnetic Field Reversal around an Active Fast Radio Burst. *arXiv* **2022**, arXiv:2203.08151.
35. Bannister, K.W.; Deller, A.T.; Phillips, C.; Macquart, J.P.; Prochaska, J.X.; Tejos, N.; Ryder, S.D.; Sadler, E.M.; Shannon, R.M.; Simha, S.; et al. A single fast radio burst localized to a massive galaxy at cosmological distance. *Science* **2019**, *365*, 565–570.
36. Ravi, V.; Catha, M.; D'Addario, L.; Djorgovski, S.G.; Hallinan, G.; Hobbs, R.; Kocz, J.; Kulkarni, S.R.; Shi, J.; Vedantham, H.K.; et al. A fast radio burst localized to a massive galaxy. *Nature* **2019**, *572*, 352–354. [CrossRef] [PubMed]
37. Marcote, B.; Nimmo, K.; Hessels, J.W.T.; Tendulkar, S.P.; Bassa, C.G.; Paragi, Z.; Keimpema, A.; Bhardwaj, M.; Karuppusamy, R.; Kaspi, V.M.; et al. A repeating fast radio burst source localized to a nearby spiral galaxy. *Nature* **2020**, *577*, 190–194.
38. Bhandari, S.; Sadler, E.M.; Prochaska, J.X.; Simha, S.; Ryder, S.D.; Marnoch, L.; Bannister, K.W.; Macquart, J.P.; Flynn, C.; Shannon, R.M.; et al. The Host Galaxies and Progenitors of Fast Radio Bursts Localized with the Australian Square Kilometre Array Pathfinder. *Astrophys. J.* **2020**, *895*, L37.
39. Heintz, K.E.; Prochaska, J.X.; Simha, S.; Platts, E.; Fong, W.f.; Tejos, N.; Ryder, S.D.; Aggerwal, K.; Bhandari, S.; Day, C.K.; et al. Host Galaxy Properties and Offset Distributions of Fast Radio Bursts: Implications for Their Progenitors. *Astrophys. J.* **2020**, *903*, 152.
40. Fong, W.f.; Dong, Y.; Leja, J.; Bhandari, S.; Day, C.K.; Deller, A.T.; Kumar, P.; Prochaska, J.X.; Scott, D.R.; Bannister, K.W.; et al. Chronicling the Host Galaxy Properties of the Remarkable Repeating FRB 20201124A. *Astrophys. J.* **2021**, *919*, L23.
41. Ravi, V.; Law, C.J.; Li, D.; Aggarwal, K.; Bhardwaj, M.; Burke-Spolaor, S.; Connor, L.; Lazio, T.J.W.; Simard, D.; Somalwar, J.; et al. The host galaxy and persistent radio counterpart of FRB 20201124A. *Mon. Not. R. Astron. Soc.* **2022**, *513*, 982–990.
42. Gordon, A.C.; Fong, W.f.; Kilpatrick, C.D.; Eftekhari, T.; Leja, J.; Prochaska, J.X.; Nugent, A.E.; Bhandari, S.; Blanchard, P.K.; Caleb, M.; et al. The Demographics, Stellar Populations, and Star Formation Histories of Fast Radio Burst Host Galaxies: Implications for the Progenitors. *arXiv* **2023**, arXiv:2302.05465. [CrossRef]
43. Li, Y.; Zhang, B. A Comparative Study of Host Galaxy Properties between Fast Radio Bursts and Stellar Transients. *Astrophys. J.* **2020**, *899*, L6.
44. Zhang, R.C.; Zhang, B. The CHIME Fast Radio Burst Population Does Not Track the Star Formation History of the Universe. *Astrophys. J.* **2022**, *924*, L14.
45. Hashimoto, T.; Goto, T.; Chen, B.H.; Ho, S.C.C.; Hsiao, T.Y.Y.; Wong, Y.H.V.; On, A.Y.L.; Kim, S.J.; Kilerci-Eser, E.; Huang, K.C.; et al. Energy functions of fast radio bursts derived from the first CHIME/FRB catalogue. *Mon. Not. R. Astron. Soc.* **2022**, *511*, 1961–1976.
46. Qiang, D.C.; Li, S.L.; Wei, H. Fast radio burst distributions consistent with the first CHIME/FRB catalog. *J. Cos. AstroPart. Phys.* **2022**, *2022*, 040.
47. Shin, K.; Masui, K.W.; Bhardwaj, M.; Cassanelli, T.; Chawla, P.; Dobbs, M.; Dong, F.A.; Fonseca, E.; Gaensler, B.M.; Herrera-Martín, A.; et al. Inferring the Energy and Distance Distributions of Fast Radio Bursts Using the First CHIME/FRB Catalog. *Astrophys. J.* **2023**, *944*, 105.
48. Bhardwaj, M.; Gaensler, B.M.; Kaspi, V.M.; Landecker, T.L.; Mckinven, R.; Michilli, D.; Pleunis, Z.; Tendulkar, S.P.; Andersen, B.C.; Boyle, P.J.; et al. A Nearby Repeating Fast Radio Burst in the Direction of M81. *Astrophys. J.* **2021**, *910*, L18.
49. Kirsten, F.; Marcote, B.; Nimmo, K.; Hessels, J.W.T.; Bhardwaj, M.; Tendulkar, S.P.; Keimpema, A.; Yang, J.; Snelders, M.P.; Scholz, P.; et al. A repeating fast radio burst source in a globular cluster. *Nature* **2022**, *602*, 585–589.
50. Nimmo, K.; Hessels, J.W.T.; Keimpema, A.; Archibald, A.M.; Cordes, J.M.; Karuppusamy, R.; Kirsten, F.; Li, D.Z.; Marcote, B.; Paragi, Z. Highly polarized microstructure from the repeating FRB 20180916B. *Nat. Astron.* **2021**, *5*, 594–603.
51. Lu, W.; Beniamini, P.; Kumar, P. Implications of a rapidly varying FRB in a globular cluster of M81. *Mon. Not. R. Astron. Soc.* **2022**, *510*, 1867–1879.
52. Kremer, K.; Li, D.; Lu, W.; Piro, A.L.; Zhang, B. Prospects for Detecting Fast Radio Bursts in the Globular Clusters of Nearby Galaxies. *Astrophys. J.* **2023**, *944*, 6.
53. Amiri, M. et al. [CHIME/FRB Collaboration] Periodic activity from a fast radio burst source. *arXiv* **2020**, arXiv:2001.10275.
54. Pastor-Marazuela, I.; Connor, L.; van Leeuwen, J.; Maan, Y.; ter Veen, S.; Bilous, A.; Oostrum, L.; Petroff, E.; Straal, S.; Vohl, D.; et al. Chromatic periodic activity down to 120 megahertz in a fast radio burst. *Nature* **2021**, *596*, 505–508.
55. Pleunis, Z.; Michilli, D.; Bassa, C.G.; Hessels, J.W.T.; Naidu, A.; Andersen, B.C.; Chawla, P.; Fonseca, E.; Gopinath, A.; Kaspi, V.M.; et al. LOFAR Detection of 110–188 MHz Emission and Frequency-dependent Activity from FRB 20180916B. *Astrophys. J.* **2021**, *911*, L3.

56. Rajwade, K.M.; Mickaliger, M.B.; Stappers, B.W.; Morello, V.; Agarwal, D.; Bassa, C.G.; Breton, R.P.; Caleb, M.; Karastergiou, A.; Keane, E.F.; et al. Possible periodic activity in the repeating FRB 121102. *Mon. Not. R. Astron. Soc.* **2020**, *495*, 3551–3558.
57. Andersen, B.C. et al. [CHIME/FRB Collaboration] Sub-second periodicity in a fast radio burst. *Nature* **2022**, *607*, 256–259.
58. Moroianu, A.; Wen, L.; James, C.W.; Ai, S.; Kovalam, M.; Panther, F.; Zhang, B. An assessment of the Association Between a Fast Radio Burst and Binary Neutron Star Merger. *arXiv* **2022**, arXiv:2212.00201.
59. Bhardwaj, M.; Palmese, A.; Magaña Hernandez, I.; D'Emilio, V.; Morisaki, S. GW190425 and FRB20190425A: Challenges for Fast Radio Bursts as Multi-Messenger Sources from Binary Neutron Star Mergers. *arXiv* **2023**, arXiv:2306.00948. [CrossRef]
60. Panther, F.H.; Anderson, G.E.; Bhandari, S.; Goodwin, A.J.; Hurley-Walker, N.; James, C.W.; Kawka, A.; Ai, S.; Kovalam, M.; Moroianu, A.; et al. The most probable host of CHIME FRB 190425A, associated with binary neutron star merger GW190425, and a late-time transient search. *arXiv* **2022**, arXiv:2212.00954.
61. Luo, R.; Wang, B.J.; Men, Y.P.; Zhang, C.F.; Jiang, J.C.; Xu, H.; Wang, W.Y.; Lee, K.J.; Han, J.L.; Zhang, B.; et al. Diverse polarization angle swings from a repeating fast radio burst source. *Nature* **2020**, *586*, 693–696.
62. Cho, H.; Macquart, J.P.; Shannon, R.M.; Deller, A.T.; Morrison, I.S.; Ekers, R.D.; Bannister, K.W.; Farah, W.; Qiu, H.; Sammons, M.W.; et al. Spectropolarimetric Analysis of FRB 181112 at Microsecond Resolution: Implications for Fast Radio Burst Emission Mechanism. *Astrophys. J.* **2020**, *891*, L38.
63. Day, C.K.; Deller, A.T.; Shannon, R.M.; Qiu, H.; Bannister, K.W.; Bhandari, S.; Ekers, R.; Flynn, C.; James, C.W.; Macquart, J.P.; et al. High time resolution and polarization properties of ASKAP-localized fast radio bursts. *Mon. Not. R. Astron. Soc.* **2020**, *497*, 3335–3350.
64. Petroff, E.; Bailes, M.; Barr, E.D.; Barsdell, B.R.; Bhat, N.D.R.; Bian, F.; Burke-Spolaor, S.; Caleb, M.; Champion, D.; Chandra, P.; et al. A real-time fast radio burst: Polarization detection and multiwavelength follow-up. *Mon. Not. R. Astron. Soc.* **2015**, *447*, 246–255.
65. Masui, K.; Lin, H.H.; Sievers, J.; Anderson, C.J.; Chang, T.C.; Chen, X.; Ganguly, A.; Jarvis, M.; Kuo, C.Y.; Li, Y.C.; et al. Dense magnetized plasma associated with a fast radio burst. *Nature* **2015**, *528*, 523–525.
66. Caleb, M.; Keane, E.F.; van Straten, W.; Kramer, M.; Macquart, J.P.; Bailes, M.; Barr, E.D.; Bhat, N.D.R.; Bhandari, S.; Burgay, M.; et al. The Survey for Pulsars and Extragalactic Radio Bursts—III. Polarization properties of FRBs 160102 and 151230. *Mon. Not. R. Astron. Soc.* **2018**, *478*, 2046–2055.
67. Kumar, P.; Shannon, R.M.; Lower, M.E.; Bhandari, S.; Deller, A.T.; Flynn, C.; Keane, E.F. Circularly polarized radio emission from the repeating fast radio burst source FRB 20201124A. *Mon. Not. R. Astron. Soc.* **2022**, *512*, 3400–3413.
68. Feng, Y.; Zhang, Y.K.; Li, D.; Yang, Y.P.; Wang, P.; Niu, C.H.; Dai, S.; Yao, J.M. Circular polarization in two active repeating fast radio bursts. *Sci. Bull.* **2022**, *67*, 2398–2401.
69. Jiang, J.C.; Wang, W.Y.; Xu, H.; Xu, J.W.; Zhang, C.F.; Wang, B.J.; Zhou, D.J.; Zhang, Y.K.; Niu, J.R.; Lee, K.J.; et al. FAST observations of an extremely active episode of FRB 20201124A: III. Polarimetry. *arXiv* **2022**, arXiv:2210.03609.
70. Hessels, J.W.T.; Spitler, L.G.; Seymour, A.D.; Cordes, J.M.; Michilli, D.; Lynch, R.S.; Gourdji, K.; Archibald, A.M.; Bassa, C.G.; Bower, G.C.; et al. FRB 121102 Bursts Show Complex Time-Frequency Structure. *Astrophys. J.* **2019**, *876*, L23.
71. Qu, Y.; Zhang, B. Polarization of fast radio bursts: Radiation mechanisms and propagation effects. *Mon. Not. R. Astron. Soc.* **2023**, *522*, 2448–2477.
72. Wang, W.Y.; Jiang, J.C.; Lee, K.; Xu, R.; Zhang, B. Polarization of magnetospheric curvature radiation in repeating fast radio bursts. *Mon. Not. R. Astron. Soc.* **2022**, *517*, 5080–5089.
73. Tong, H.; Wang, H.G. Circular polarisation of fast radio bursts in the curvature radiation scenario. *arXiv* **2022**, arXiv:2202.05475.
74. Kumar, P.; Lu, W.; Bhattacharya, M. Fast radio burst source properties and curvature radiation model. *Mon. Not. R. Astron. Soc.* **2017**, *468*, 2726–2739.
75. Yang, Y.P.; Zhang, B. Bunching Coherent Curvature Radiation in Three-dimensional Magnetic Field Geometry: Application to Pulsars and Fast Radio Bursts. *Astrophys. J.* **2018**, *868*, 31.
76. Zhang, B. Coherent Inverse Compton Scattering by Bunches in Fast Radio Bursts. *Astrophys. J.* **2022**, *925*, 53.
77. Beniamini, P.; Kumar, P. What does FRB light-curve variability tell us about the emission mechanism? *Mon. Not. R. Astron. Soc.* **2020**, *498*, 651–664.
78. Yang, Y.P. Implications of Spectra and Polarizations of Fast Radio Bursts: From Perspective of Radiation Mechanisms. *arXiv* **2023**, arXiv:2305.08649. [CrossRef]
79. Sironi, L.; Plotnikov, I.; Nättilä, J.; Beloborodov, A.M. Coherent Electromagnetic Emission from Relativistic Magnetized Shocks. *Phys. Rev. Lett.* **2021**, *127*, 035101.
80. Szary, A.; Zhang, B.; Melikidze, G.I.; Gil, J.; Xu, R.X. Radio Efficiency of Pulsars. *Astrophys. J.* **2014**, *784*, 59. [CrossRef]
81. Wang, W.; Zhang, B.; Chen, X.; Xu, R. On the Time-Frequency Downward Drifting of Repeating Fast Radio Bursts. *Astrophys. J.* **2019**, *876*, L15. [CrossRef]
82. Lyutikov, M. Radius-to-frequency Mapping and FRB Frequency Drifts. *Astrophys. J.* **2020**, *889*, 135. [CrossRef]
83. Metzger, B.D.; Margalit, B.; Sironi, L. Fast radio bursts as synchrotron maser emission from decelerating relativistic blast waves. *Mon. Not. R. Astron. Soc.* **2019**, *485*, 4091–4106.
84. Kundu, E.; Zhang, B. Free-free absorption in hot relativistic flows: Application to fast radio bursts. *Mon. Not. R. Astron. Soc.* **2021**, *508*, L48–L52.
85. Lyubarsky, Y. Emission Mechanisms of Fast Radio Bursts. *Universe* **2021**, *7*, 56. [CrossRef]

86. Qu, Y.; Zhang, B.; Kumar, P. The plasma suppression effect can be ignored in realistic FRB models invoking bunched coherent radio emission. *Mon. Not. R. Astron. Soc.* **2023**, *518*, 66–74.
87. Wadiasingh, Z.; Timokhin, A. Repeating Fast Radio Bursts from Magnetars with Low Magnetospheric Twist. *Astrophys. J.* **2019**, *879*, 4.

Article

Galaxy Rotation Curve Fitting Using Machine Learning Tools

Carlos R. Argüelles [1,2,*,†] and Santiago Collazo [1,*,†]

[1] Instituto de Astrofísica de La Plata, UNLP-CONICET, Paseo del Bosque s/n, La Plata B1900FWA, Argentina
[2] ICRANet, Piazza della Repubblica 10, I-65122 Pescara, Italy
* Correspondence: carguelles@fcaglp.unlp.edu.ar (C.R.A.); scollazo@fcaglp.unlp.edu.ar (S.C.)
† These authors contributed equally to this work.

Abstract: Galaxy rotation curve (RC) fitting is an important technique which allows the placement of constraints on different kinds of dark matter (DM) halo models. In the case of non-phenomenological DM profiles with no analytic expressions, the art of finding RC best-fits including the full baryonic + DM free parameters can be difficult and time-consuming. In the present work, we use a gradient descent method used in the backpropagation process of training a neural network, to fit the so-called Grand Rotation Curve of the Milky Way (MW) ranging from ~ 1 pc all the way to $\sim 10^5$ pc. We model the mass distribution of our Galaxy including a bulge (inner + main), a disk, and a fermionic dark matter (DM) halo known as the Ruffini-Argüelles-Rueda (RAR) model. This is a semi-analytical model built from first-principle physics such as (quantum) statistical mechanics and thermodynamics, whose more general density profile has a *dense core–diluted halo* morphology with no analytic expression. As shown recently and further verified here, the dark and compact fermion-core can work as an alternative to the central black hole in SgrA* when including data at milliparsec scales from the S-cluster stars. Thus, we show the ability of this state-of-the-art machine learning tool in providing the best-fit parameters to the overall MW RC in the 10^{-2}–10^5 pc range, in a few hours of CPU time.

Keywords: dark matter; Milky Way; rotation curves; numerical methods

Citation: Argüelles, C.R.; Collazo S. Galaxy Rotation Curve Fitting Using Machine Learning Tools. *Universe* 2023, 9, 372. https://doi.org/10.3390/universe9080372

Academic Editors: Stephen J. Curran and Nicola R. Napolitano

Received: 5 July 2023
Revised: 10 August 2023
Accepted: 14 August 2023
Published: 16 August 2023

1. Introduction

Disk galaxies, like our own, are rotational supported structures with the advantage of having baryonic (or luminous) mass tracers in approximate circular orbits from which it is possible to obtain the so-called RC. The specific DM distribution, usually dubbed as the DM density profile, is inferred by fitting the observed velocity RC as a function of the galactocentric radius. Typically, this is carried out by assuming a given underlying DM profile together with different mass models for the luminous components such as the bulge, disc, etc. (see, e.g., [1] for a review). Most of these studies assume phenomenological DM profiles (in spherical symmetry) obtained from classical N-body cosmological simulations with a given analytic expression, besides the visible mass components. However, other kinds of DM profiles can be obtained from first-principle physics (i.e., thermodynamics and statistical mechanics) while accounting for the quantum nature of the particles, such as the RAR model for fermions (see [2] for a review, and references therein) and, e.g., [3,4] for bosonic DM, with no analytic expressions for the profiles.

The RAR model consist in a self-gravitating system of fermions in General Relativity, and therefore is built upon a coupled system of (ordinary) highly non-linear differential equations, which defines a boundary condition problem to be solved numerically (see, e.g., [5] for its original version and [6] for its more realistic extension including for particle evaporation). In its extended version, the RAR model involves four free parameters: m the particle mass, and the set (β, θ, W) of dimensionless parameters reading for the temperature, degeneracy, and cut-off particle energy, respectively. These parameters are present in the underlying coarse-grained distribution function (DF) of the particles (which is of Fermi–Dirac type as explained in Section 2.2), and have to be set at the center of the configuration (denoted with the subscript 0) in order to solve the system of equilibrium differential equations of the RAR model (see Equations (9)–(13) in Argüelles and et al. [6]).

When applied to real galaxies, in the recent past the RAR equations were solved for given boundary halo conditions taken from observations. For example, when applied to the MW in [6], three boundary conditions were considered for the overall DM halo mass: one at the fermion-core ($M_c = 4.2 \times 10^6 M_\odot$, in order to be an alternative to the BH in SgrA*), and the other two at mid-outer DM halo (i.e., $M(r = 12 \text{ kpc}) \approx 5 \times 10^{10} M_\odot$ [7], and $M(r = 40 \text{ kpc}) \approx 2 \times 10^{11} M_\odot$ [8], respectively). Since the RAR model has four free-parameters, once the particle was fixed within the range[1] $(48, 345)$ keV, there exists one *core-halo* solution for such a particle mass (i.e., three boundary conditions for three remaining free-parameters) fulfilling with the constraints. Interestingly, even if only three boundary conditions were used from observations at very different scales, the overall behavior of such a RAR DM solution is good enough to fit within the error bars of the Grand RC [6] (after standard baryonic mass models are included).

A more refined phenomenological analysis of the relativistic RAR model would require a best-fit procedure using the full data points of the corresponding RC including their errors (e.g., using MCMC or grid-coverage methods). Such kind of analysis has recently been performed in [9] within an MCMC method for a large sample of 120 galaxies of the SPARC catalog [10], with explicit χ^2 minimization and corresponding posteriors for the RAR model parameters for a fixed particle mass fixed at $m = 50$ keV. However, besides the fixed m case analyzed in [9], the baryonic mass models were fixed according to the SPARC-catalog (for each galaxy). Thus, it is of interest to develop a numerical technique which, for non-analytic DM models such as RAR, makes it possible to provide best-fits to the full RC-data when including for a larger free parameter-space (i.e., full DM + baryonic model parameters) in a few hours of CPU-time.

Thus, in this work we propose a new RC best-fitting method based on state of-the-art machine learning tools, when including for baryonic free parameters together with the full four free-RAR model parameters. In particular we will focus our attention in the so-called Grand RC as studied in [1] further including for with innermost data coming from the S-cluster stars orbiting SgrA*, thus covering in total about seven orders of magnitude in galactocentric-radius (e.g., within $\sim 10^{-2}$–10^5 pc). All in all, the ability of the RAR model to provide excellent fits to the MW RC covering very different radial-scales is shown, involving large order-of-magnitude variations in the gravitational potential generated by the overall total mass distribution of the Galaxy.

2. Rotation Curve Data and Methodology

In this section we describe the data selection together with the methodology. We first detail the different data points considered in this work, both coming from the S-cluster stars around SgrA* as analyzed in [6] from data taken in [11], and the ones coming from Pop. I stars and interstellar gas as compiled in [1] including for different observational techniques. Then we make explicit the different mass models here assumed, accounting for a central dark compact object (no-BH), inner + main bulge, disc, and outer DM halo. It is important to emphasize once more that both the massive dark compact object together with the outer halo are different components of the very same fermionic RAR DM-model; that is, the dense fermion-core (supported against gravity by fermion-degeneracy pressure) is surrounded by a more dilute DM halo (supported against gravity by thermal pressure). Finally, we briefly describe the machine learning tool methods applied here to make the best fit for the given Grand RC.

2.1. Data Selection

We use the observed Grand RC of the Milky Way as provided in [1,7], including for different combined observational methods (with associated systematics), ranging from ~ 1 pc up to $\sim 10^5$ pc. Due to large error bars arising above ≈ 40 kpc, we will consider here the overall (processed) RC data[2] up to that radial scale (see Figure 1). This RC uses, as Galactic constant values, $R_0 = 8$ kpc and $V_0 = 238$ km/s (with R_0 the distance of the sun from the Galaxy center and V_0 the circular velocity of the Local Standard of Rest (LSR) at the Sun). Additionally, we follow the analysis of the eight best-resolved S-cluster stars [11] as used in [6] by considering their average circular-orbit velocity. As clearly shown in [6], they follow the expected Keplerian velocity trend as a function of galactocentric radius due to the gravitational potential of the dense central object, all the way up to $\sim 10^{-1}$ pc [1,7]. Thus for convenience we will choose a typical (average) circular-velocity value within such an inner Keplerian trend at 10^{-2} pc (see innermost data point in Figure 1). The combination of such different radial scales covered in the data here selected is, of course, motivated in view of the *core-halo* nature of the RAR DM model, for which a best-fit in their free parameters will be attempted.

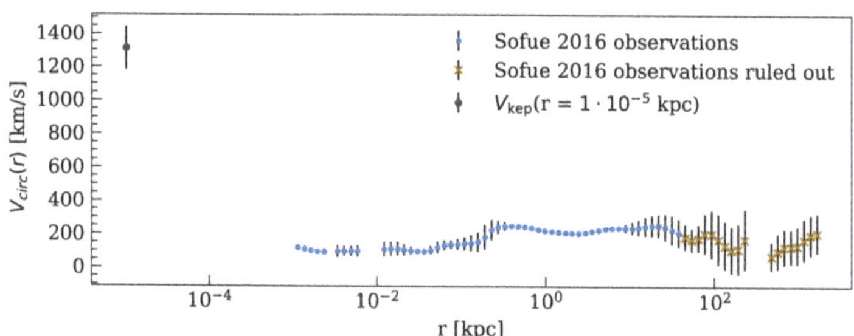

Figure 1. Overall Milky Way RC used to constrain the gravitational potential of the Galaxy. It is composed by the Grand Rotation Curve from [1] (light-blue and orange dots) and an inner (milliparsec scale) keplerian velocity (grey dot) of an S-cluster star as caused by the central object in SgrA*.

2.2. Mass Models of the MW

The full dynamical parameters for an accurate Galactic mass determination is very complex. As detailed in [1], they can be divided into three main categories: an axisymmetric structure (or RC); a non-axisymmetric structure (outside the RC) including inner bars and arms; and a radial flow (out of the RC) including ring structures. The work of [1] centered the attention in the RC only, as is also the case of the present paper. This axisymmetric structure includes a dark central object, a bulge (with two components), a flat disc and an spherical DM halo. Thus, we model the gravitational potential of the Milky Way through the following different components:

(i) The bulge is going to be modeled by two exponential spheroids according to [1,7], one to model the inner bulge and the other to model the main bulge. The matter density for such models, each one providing two free parameters, is:

$$\rho(r) = \rho_c \exp\left(-r/a_b\right), \tag{1}$$

where $a_{b(i)} = 3.5 \times 10^{-3}$ kpc; $a_{b(m)} = 1.2 \times 10^{-1}$ kpc; $\rho_{c(i)} = 3.7 \times 10^{13} M_\odot/\text{kpc}^3$ and $\rho_{c(m)} = 2.1 \times 10^{11} M_\odot/\text{kbulgebulgepc}^3$, with the sub-indices i and m indicating the inner and main components. While the bulge parameters will be kept fixed in this work for definitness, the free parameters of the disk will be varied along the RAR model parameters to provide the best fit.

(ii) The disk is going to be modeled by an exponential flat disk as studied in [7]. The surface mass density of such a disc provides for two free parameters (Σ_d, a_d), and reads

$$\Sigma(R) = \Sigma_d \exp\left(-R/a_d\right), \tag{2}$$

where R is the standard cylindrical radius, and (Σ_d, a_d) to be determined by our best fitting procedure.

(iii) Both the dark central compact object together with the DM halo will be modeled by the semi-analytical (extended) RAR model, which, as explained in the Introduction, is based on a self-gravitating system of fermions at finite temperature including for escape of particles and central fermion-degeneracy. This DM model was extensively studied in [2,6] and references therein for the Milky Way, and in [9] for other galaxy types. It has four free-parameters $(m, \theta_0, W_0, \beta_0)$ with m the DM, particle mass, and (β_0, θ_0, W_0) the dimensionless parameters evaluated at the origin, reading for the temperature, degeneracy, and cut-off particle energy, respectively. The free-RAR model parameters enter in the underlying phase-space DF of the fermions at (quasi) equilibrium, whose formula is given in Equation (3) below. Interestingly, it can be demonstrated [12,13] that such a Fermi–Dirac-like DF is a quasi-stationary solution of a kinetic theory equation (of Fermionic–Landau form) via the application of a maximum entropy principle. Thus it is a most-probable coarse-grained DF at violent relaxation, extending the original results of Lynden–Bell on the subject.

$$\bar{f}(\epsilon \le \epsilon_c) = \frac{1 - e^{(\epsilon - \epsilon_c)/kT}}{e^{(\epsilon - \mu)/kT} + 1}, \qquad \bar{f}(\epsilon > \epsilon_c) = 0, \tag{3}$$

where $\epsilon = \sqrt{c^2 p^2 + m^2 c^4} - mc^2$ is the particle kinetic energy, μ is the chemical potential with the particle rest-energy subtracted off, T is the effective temperature, k is the Boltzmann constant, c is the speed of light, and m is the DM fermion mass. The already defined set of dimensionless-parameters are $\beta = kT/(mc^2)$, $\theta = \mu/(kT)$ and $W = \epsilon_c/(kT)$, respectively. It has been further shown [14] that DM halos built upon such a fermionic DF can be done within a Warm DM cosmological framework, and naturally leading to stable halos which can be extremely long-lived with key implications to the formation and further growth of supermassive BHs in the early Universe [2].

This kind of system, in its most general morphology, develops a density profile with a *dense core–diluted halo* distribution (see Figure 2), and is supported against gravity through Fermi degeneracy pressure (for the core) and by thermal pressure (in the outer halo).

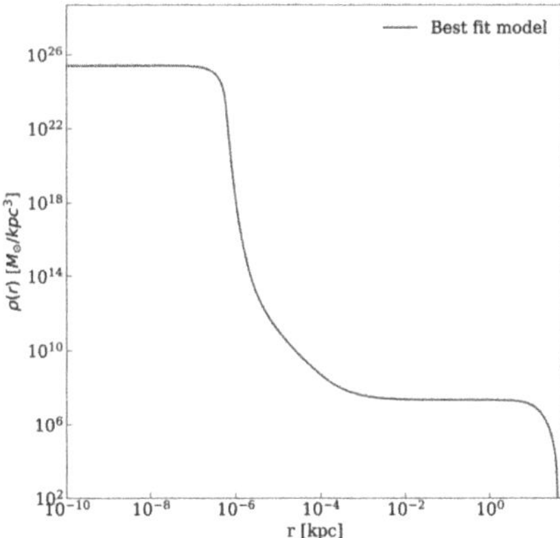

Figure 2. Density profile of the RAR model constrained in this work by the gradient descent method explained in the main text. It can be seen the *constant-density* core below the mili-pc scale (which is governed by quantum degeneracy pressure) and the transition to the plateau at $\sim$1 pc (where quantum effects are negligible), while far above $\sim$1 kpc, it follows a polytropic tail (see also [9]).

2.3. Gradient Descent Method: A Machine Learning Tool

The gradient descent method is based on a progressive sequence of steps to minimize a function. Given a function F to be minimized, it will implement the formula

$$\mathbf{p^{new}} = \mathbf{p^{old}} - \gamma \nabla_{\mathbf{p}} F(\mathbf{p^{old}}) \tag{4}$$

where **p** stands for the independent variable of the function F and γ is a parameter called learning rate whose aim is to regulate the "length" of the steps. If this formula is implemented recursively, one can eventually go closer and closer to a minimum of F. An illustration of the procedure followed by the gradient descent method is shown in Figure 3. In that image can be seen a solid dark path, which is the result of evaluating F in the different points given by Equation (4). Since such a formula uses "minus" the gradient of the function, the path followed by the method is oriented to the direction of "maximum" decreasing, driving to the deepest point of F.

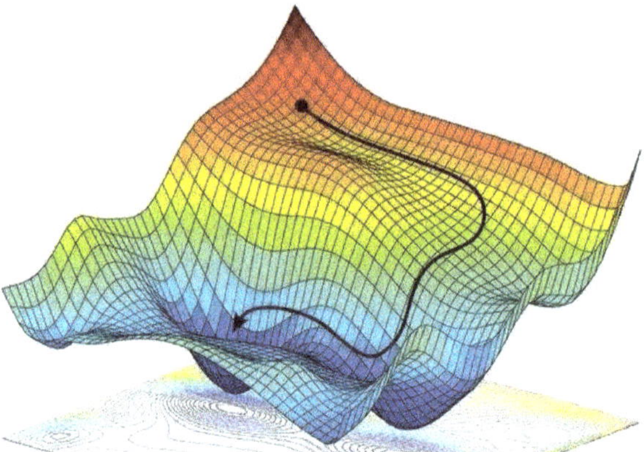

Figure 3. Illustration of the path followed by the gradient descent method to reach the minimum of a two-dimensional function. Taken from: https://easyai.tech/en/ai-definition/gradient-descent/ (accessed on 10 May 2023).

In the following we provide some pros and cons when using this algorithm. The main advantage is that it finds the minimum of F in a more direct manner (and usually more precisely) than an MCMC or grid-coverage methods: it does not need to "explore" a huge volume of different values of F to see which is the minimum one. Instead, it starts to *walk* in a direction and, after iterating the steps, it directly goes towards the minimum. Because of this, it can happen that the algorithm becomes stuck in a local minimum rather than in the *global* one. This is a disadvantage of the method and there are some techniques to avoid this local minima. In our case, we chose to use the gradient descent method as a fine-tuning algorithm of a method providing a "partial" minimum of F. That is to say, we took as the initial seed for the gradient descent method the final result of another general optimization method—that is a genetic algorithm as implemented in the optimized RAR code in PYTHON which will be publicly available through Github in 2023—the latter making a faster (though less accurate) exploration of the parameter space.

It is important to emphasize that the idea of this paper is to present a neural-network-prepared tool so it can fit a non-linear non-analytic model with several free parameters like the one here shown. It was not our intention to perform a deep analysis of the fitting technique and/or give a detailed comparison with other tools. So, since there is no ideal way of dealing with the local minima problem, also in neural networks, what we have planned was to apply this method to perform a 'fine-tuning' of the fitting problem once a good-enough initial parameters-seed is provided by other commonly used method (e.g., a genetic algorithm). Thus this fine-tuning technique can be used by anyone who intends to improve the accuracy in the fitted parameters, in highly non-linear models such as, for example, those involving General Relativistic Einstein equations as in the RAR model. With respect to the time savings, the most important comparison is between this method and a traditional MCMC fitting approach. While this method took us 1.5 or 2 h to complete the fine-tuning of the free parameters based on a good-fit seed of them, the MCMC approach we have tried before took us several hours or even more (depending on the chains), and only for two free parameters; in this case, we could fit six free parameters.

Another important comment has to do with a useful advantage related to the morphology of the gradient descent method, since it resides on computing a gradient, and therefore it is less sensitive to the dimension of the parameter space than the MCMC or grid-coverage methods. The latter methods have to walk around all over the subspace of parameters, increasing their computing time drastically for high-dimensional spaces. Indeed, while a very precise RC best-fit for the Milky Way under the same RAR model as used here (with varying baryonic parameters) can take more than a day with genetic algorithms (similar to MCMC), the method implemented here takes a few hours of CPU-time.

In the case of this work, the function F introduced above will be a function that quantifies how good the predictions made by the model are in contrast to the observations. In machine learning, these kinds of functions are called *loss functions*. Specifically, we will use a Mean Squared Error (MSE) as a loss function, which is defined as:

$$Loss(\mathbf{p}) = 1/C \sum_{i=1}^{N} \frac{(V(r_i, \mathbf{p}) - v_i)^2}{N}, \tag{5}$$

where $\mathbf{p}$ is the vector of the (physical) free parameters that characterize the full mass model (baryons + DM). In this work we fit six free parameters (four in the RAR DM halo plus two of the Freeman disk), adopting the bulge free parameters as detailed in Section 2.2. The predicted circular velocities of the different mass models are denoted with $V(r_i, \mathbf{p})$ and v_i are the observed ones, C is a normalization constant and N is the number of observations. The idea is to fit the free parameters of the model mentioned above to the overall rotation curve (milliparsec inner point + Grand RC).

The implementation of the algorithm was carried out with the help of a tool provided by the PyTorch package [15], which is an open-source machine learning framework. Such a package was naturally incorporated in the RAR model's code, the last version of which was programmed in PYTHON language (a publicly available version will be published in 2023 as an open code via Github). Since it allows the possibility to include neural networks, it has incorporated the gradient descent method widely used during the training process of such machine learning algorithms. Instead of defining a neural network, we used the backpropagation numerical method, which is able to compute derivatives of compound functions. The repeated application of this numerical method has allowed us to apply the gradient descent method to this specific problem.

3. Results

In this section, we briefly present the best-fit results to the RC of our Galaxy. The model was iterated with a learning rate of $\gamma = 0.001$ and through 1500 epochs (see Figure 4), giving the following fit (see Figure 5) to the so-called Grand Milky Way RC as given in [1]. That is, it covers from inner bulge data points (starting at few pc), to the main bulge data points (peaking at about 0.4 kpc), to data points throughout the disk and DM halo region up to several tens of kpc, all including error bars. In addition, we include independent data points to those analyzed in [1], reaching down to milliparsec scales, and coming from the best resolved S-cluster stars orbiting SgrA* [16].

In the case of Figure 5 and for the sake of simplicity, we have included for the S-stars a single data point located at 0.01 pc, corresponding to the average circular velocity of a typical S-cluster star which falls along the Keplerian velocity trend caused by the supermassive central object. It is important to remark that the behaviour of the circular RC of the Milky Way as predicted by the RAR model (without the central BH) between the innermost data point for a typical S-cluster star at 0.01 pc and the first data point provided in [1], is Keplerian (i.e., $V_{circ} \propto r^{-1/2}$) all the way to the radius of gravitational influence of the central object at ~ 1 pc, as expected (see also [6]) for an analogous plot including for the eight best-resolved S-cluster stars).

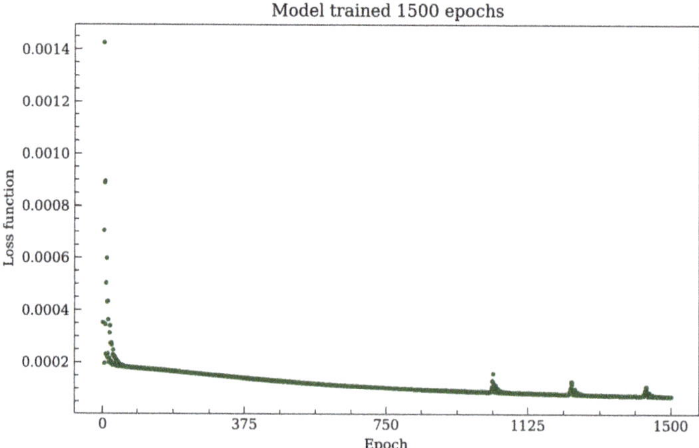

Figure 4. Loss function against the step or *epochs* of the method. It can be seen that it is a decreasing function, with little bumps at the tail on the right, indicating a clear minimization trend reaching the value of 0.000071979 after 1500 epochs.

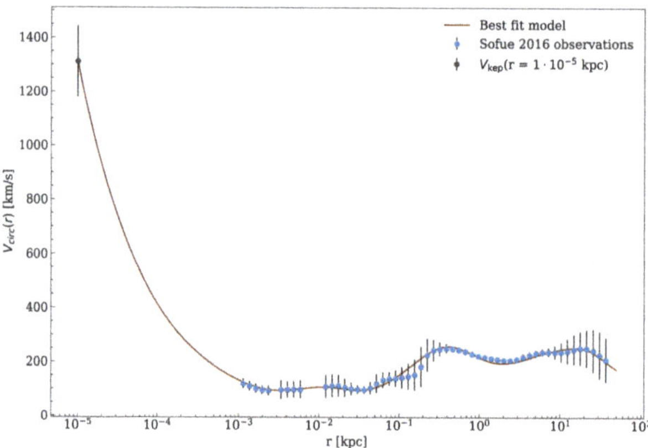

Figure 5. Best-fit circular velocity curve result of the implementation of the gradient descent method. It is remarkable the very good precision achieved in almost all the data-points in few hours CPU time, despite some minor deficiencies at the $\sim 10^{-1}$ kpc scale.

It can be seen above that the model generates an excellent fit throughout the whole Galaxy range. The loss value at the last training epoch is 0.000071979 (see Figure 4). The full best-fit parameters obtained from the iterated model and are given in Table 1 below.

Table 1. Best-fit parameters results after applying the gradient descent method to the set of Milky Way observables under the gravitational potential model described in the main text.

Parameter	Seed Value	Final Value
m [keV/c^2]	56.0	54.809
θ_0	37.766	37.809
W_0	66.341	66.449
β_0	1.1977×10^{-5}	1.1139×10^{-5}
Σ_d [$M_\odot$/kpc^2]	5.9658×10^8	1.0882×10^9
a_d [kpc]	4.9	3.0039

From the above Table we can see that the final (best-fit) values of the disc parameters are in line with those reported in [1,7], though the resulting disc here is somewhat more massive with differences of up to $\sim$25% or $\sim$50% in the surface DM density parameter respectively. Such differences in the barionic free parameters are expected, since in [1,7] it was assumed to have a NFW profile which is of a cuspy nature (the RAR is cored) and the RAR outer density tail is politropic while the NFW one is a power law (see [9] for an extensive comparison of possible RAR density morphologies with respect to other typical DM density profiles considered in the literature).

Regarding the overall DM distribution, the best-fit DM free-parameters imply a compact fermion-core of mass $M_c = 3.46 \times 10^6 M_\odot$ and radius $r_c = 4.54 \times 10^{-4}$ pc (see Figure 2), thus well within the pericenter of the closest and best-resolved S-cluster star: the S-2 star, which has indeed served as the best (and most accurate) case to constrain the mass of the supermassive dark compact object in SgrA* [11]. This is totally in line with the recent results reviewed in [2] within a different phenomenological approach (i.e., fixing only three boundary halo mass conditions from observations as explained in Section 1), where the ability of the RAR model to explain the astrometric data of the S-cluster stars, including the relativistic effects (gravitational redshift and orbit precession) of the S-2 star, was explicitly demonstrated. For the total mass of the Galaxy, we obtain from this work $M_{tot} \approx 3.4 \times 10^{11} M_\odot$ in line with the results obtained in [8] and in [6] (see footnote 1 in the latter for further discussion). Finally, we report a local DM density at the Sun (i.e., at $R_0 = 8$ kpc) of 0.53 GeV/cm^3, which is well within the 2σ value as reported in [17] for another cored (i.e., Burkert) DM profile.

4. Discussion and Conclusions

We have implemented a state-of-the-art machine learning numerical technique to best-fit an RC of disc galaxies which include non-analytic DM density profiles. This numerical tool uses a gradient descent method used in the backpropagation process of training a neural network, which can easily include several free-parameters and provide a best-fit within a few hours of CPU-time, thus improving (in some aspects) on other best-fitting methods. As an example study, we have chosen the well-investigated case of our own Galaxy by Y. Sofue [1] within the so-called Grand Rotation Curve, and we have further included for milliparsec velocity data coming from the S-cluster star orbiting SgrA* (thus covering a wide range of scales from 10^{-2} pc all the way to 10^5 pc). A key advantage of the machine learning tool used here (i.e., the gradient descent method implemented through Pytorch) is that it can achieve an excellent accuracy in best-fitting an RC under no-analytic DM halo models, such as the RAR halo model, together with baryonic mass models with varying free parameters within a few hours' time. The relevance of using this kind of DM profile instead of those others commonly used in the literature relies on the fact it is a semi-analytical model built from first-principle physics, such as (quantum) statistical mechanics and thermodynamics, whose more general density profile has a *dense core–diluted halo* which depends on the DM particle mass. As recently shown in [2,6] and references therein, and further verified here, the dark and compact fermion-core can work as an alternative to the central black hole in SgrA* when including data at milliparsec scales from the S-cluster stars.

Author Contributions: Conceptualization, C.R.A. and S.C.; methodology, C.R.A. and S.C.; software, S.C.; validation, C.R.A.; formal analysis, S.C.; investigation, C.R.A. and S.C.; resources, C.R.A.; data curation, S.C.; writing—original draft preparation, C.R.A.; writing—review and editing, C.R.A. and S.C.; visualization, S.C.; supervision, C.R.A.; project administration, C.R.A.; funding acquisition, C.R.A. All authors have read and agreed to the published version of the manuscript.

Funding: C.R.A. and S.C. were supported by CONICET of Argentina, and the ANPCyT (grant PICT-2018-03743).

Data Availability Statement: The dataset used in this work for the Grand RC was taken from the Y. Sofue's public database: http://www.ioa.s.u-tokyo.ac.jp/~sofue/htdocs/2017paReview/ (accessed on 15 July 2022) while the innermost central data point at milliparsec scale was taken from the analysis of the S-cluster stars (average circular velocity calculation) as done in [6].

Acknowledgments: C.R.A. acknowledges support from ICRANet.

Conflicts of Interest: The authors declare no conflict of interest.

Notes

1. The compactness of the fermion-core is inversely proportional to m [6], and thus it is shown that for $m < 48$ keV the core is too extended to fit within the S-2 star pericenter, while for $m > 345$ keV the solutions are unstable since the critical value for collapse to a BH is reached at $m = 345$ keV.
2. http://www.ioa.s.u-tokyo.ac.jp/~sofue/htdocs/2017paReview/ (accessed on 15 July 2022).

References

1. Sofue, Y. Rotation and mass in the Milky Way and spiral galaxies. *Publ. Astron. Soc. Jpn.* **2017**, *69*, R1. [CrossRef]
2. Argüelles, C.R.; Becerra-Vergara, E.A.; Rueda, J.A.; Ruffini, R. Fermionic Dark Matter: Physics, Astrophysics, and Cosmology. *Universe* **2023**, *9*, 197. [CrossRef]
3. Bernal, T.; Fernández-Hernández, L.M.; Matos, T.; Rodríguez-Meza, M.A. Rotation curves of high-resolution LSB and SPARC galaxies with fuzzy and multistate (ultralight boson) scalar field dark matter. *Mon. Not. R. Astron. Soc.* **2018**, *475*, 1447–1468. [CrossRef]
4. Robles, V.H.; Bullock, J.S.; Boylan-Kolchin, M. Scalar field dark matter: Helping or hurting small-scale problems in cosmology? *Mon. Not. R. Astron. Soc.* **2019**, *483*, 289–298. [CrossRef]
5. Ruffini, R.; Argüelles, C.R.; Rueda, J.A. On the core-halo distribution of dark matter in galaxies. *Mon. Not. R. Astron. Soc.* **2015**, *451*, 622–628. [CrossRef]
6. Argüelles, C.R.; Krut, A.; Rueda, J.A.; Ruffini, R. Novel constraints on fermionic dark matter from galactic observables I: The Milky Way. *Phys. Dark Universe* **2018**, *21*, 82–89. [CrossRef]
7. Sofue, Y. Rotation Curve and Mass Distribution in the Galactic Center - From Black Hole to Entire Galaxy. *Publ. Astron. Soc. Jpn.* **2013**, *65*, 118. [CrossRef]
8. Gibbons, S.L.J.; Belokurov, V.; Evans, N.W. 'Skinny Milky Way please', says Sagittarius. *Mon. Not. R. Astron. Soc.* **2014**, *445*, 3788–3802. [CrossRef]
9. Krut, A.; Argüelles, C.R.; Chavanis, P.H.; Rueda, J.A.; Ruffini, R. Galaxy Rotation Curves and Universal Scaling Relations: Comparison between Phenomenological and Fermionic Dark Matter Profiles. *Astrophys. J.* **2023**, *945*, 1. [CrossRef]
10. Lelli, F.; McGaugh, S.S.; Schombert, J.M. SPARC: Mass Models for 175 Disk Galaxies with Spitzer Photometry and Accurate Rotation Curves. *Astron. J.* **2016**, *152*, 157. [CrossRef]
11. Gillessen, S.; Eisenhauer, F.; Fritz, T.K.; Bartko, H.; Dodds-Eden, K.; Pfuhl, O.; Ott, T.; Genzel, R. The Orbit of the Star S2 Around SGR A* from Very Large Telescope and Keck Data. *Astrophys. J.* **2009**, *707*, L114–L117. [CrossRef]
12. Chavanis, P.H. On the 'coarse-grained' evolution of collisionless stellar systems. *Mon. Not. R. Astron. Soc.* **1998**, *300*, 981–991. [CrossRef]
13. Chavanis, P.H. Generalized thermodynamics and kinetic equations: Boltzmann, Landau, Kramers and Smoluchowski. *Phys. A Stat. Mech. Appl.* **2004**, *332*, 89–122. [CrossRef]
14. Argüelles, C.R.; Díaz, M.I.; Krut, A.; Yunis, R. On the formation and stability of fermionic dark matter haloes in a cosmological framework. *Mon. Not. R. Astron. Soc.* **2021**, *502*, 4227–4246. [CrossRef]
15. Paszke, A.; Gross, S.; Massa, F.; Lerer, A.; Bradbury, J.; Chanan, G.; Killeen, T.; Lin, Z.; Gimelshein, N.; Antiga, L.; et al. PyTorch: An Imperative Style, High-Performance Deep Learning Library. *arXiv* **2019**, arXiv:1912.01703. [CrossRef]
16. Gillessen, S.; Plewa, P.M.; Eisenhauer, F.; Sari, R.; Waisberg, I.; Habibi, M.; Pfuhl, O.; George, E.; Dexter, J.; von Fellenberg, S.; et al. An Update on Monitoring Stellar Orbits in the Galactic Center. *Astrophys. J.* **2017**, *837*, 30. [CrossRef]
17. Nesti, F.; Salucci, P. The Dark Matter halo of the Milky Way, AD 2013. *J. Cosmol. Astropart. Phys.* **2013**, *2013*, 16. [CrossRef]

universe

MDPI

Article

Estimates of the Surface Magnetic Field Strength of Radio Pulsars

Vitaliy Kim [1,2,*], Adel Umirbayeva [1,3,†] and Yerlan Aimuratov [1,3,4,*,‡]

1 Fesenkov Astrophysical Institute, Observatory 23, Almaty 050020, Kazakhstan
2 Pulkovo Observatory, Pulkovoskoe Shosse 65/1, 196140 Saint Petersburg, Russia
3 Department of Theoretical and Nuclear Physics, Faculty of Physics and Technology, Al-Farabi Kazakh National University, Al-Farabi Avenue 71, Almaty 050040, Kazakhstan
4 International Center for Relativistic Astrophysics Network, Piazza della Repubblica 10, 65122 Pescara, Italy
* Correspondence: kim@fai.kz (V.K.); aimuratov@fai.kz (Y.A.)
† Currently a Master student at Al-Farabi KazNU.
‡ Currently a Postdoctoral associate at Al-Farabi KazNU.

Abstract: We investigate the geometry of the magnetic field of rotation-powered pulsars. A new method for calculating an angle (β) between the spin and magnetic dipole axes of a neutron star (NS) in the ejector stage is considered within the frame of the magnetic dipole energy loss mechanism. We estimate the surface magnetic field strength (B_{ns}) for a population of known neutron stars in the radio pulsar (ejector) stage. The evaluated $B_{\mathrm{ns}}(\beta)$ may differ by an order of magnitude from the values without considering the angle β. It is shown that $B_{\mathrm{ns}}(\beta)$ lies in the range 10^8–10^{14} G for a known population of short and middle periodic radio pulsars.

Keywords: neutron star; radio pulsar; magnetic dipole radiation; magnetic field

Citation: Kim, V.; Umirbayeva, A.; Aimuratov, Y. Estimates of the Surface Magnetic Field Strength of Radio Pulsars. *Universe* 2023, 9, 334. https://doi.org/10.3390/universe9070334

Academic Editor: Lorenzo Iorio

Received: 2 June 2023
Revised: 12 July 2023
Accepted: 13 July 2023
Published: 14 July 2023
Corrected: 18 June 2024

1. Introduction

Radio pulsars are fast-spinning magnetized neutron stars (NS) demonstrating regular modulations (pulsations) of their radiation with a high stable period in the radio range. The axis of the magnetic field of the radio pulsar and its spin axis are not aligned, and the beam of radiation is emitted in a cone-shaped region (see Figure 1). Therefore, pulsar radiation is seen as pulses (beacon effect) by an external observer [1].

Radio pulsars are characterized by the rapid axial rotation (or spin) they have acquired due to the conservation of angular momentum during their formation. Their spin periods (P_s) lie in a wide range: from 0.0014 s to 23.5 s [2,3], with the majority not exceeding a few seconds. Radio pulsars are usually divided into several groups depending on their spin period [4,5]:

1. Short-periodic pulsars, including millisecond pulsars ($P_s < 0.1\,\mathrm{s}$);
2. Middle-periodic pulsars ($0.1\,\mathrm{s} < P_s < 2\,\mathrm{s}$);
3. Long-periodic pulsars ($P_s > 2\,\mathrm{s}$).

According to [5], a pulsar wind mainly causes the spin-down process of long-periodic radio pulsars. However, for short and middle periodic radio pulsars, a primary mechanism of rotation energy loss ($\dot{E}_{\mathrm{obs}}$) is believed to be magnetic dipole radiation (MDR).

The MDR mechanism of the energy loss was first considered in [6,7] for radio pulsars. It was shown that the magnetized NS could lose its rotational energy by MDR generation. This evolution stage of NS is also known as the "ejector" stage, and its energy loss ($\dot{E}_{\mathrm{md}}$) for the generation of MDR expresses as:

$$\dot{E}_{\mathrm{md}} = -\frac{2}{3}\frac{\mu_{\mathrm{ns}}^2 \, \omega_s^4 \, (\sin\beta)^2}{c^3}\,, \tag{1}$$

where $\mu_{\text{ns}} = B_{\text{ns}} R_{\text{ns}}^3 / 2$ is the magnetic dipole moment of NS, R_{ns} is the radius of NS, $\omega_{\text{s}} = 2\pi / P_{\text{s}}$ is the angular rotation velocity, c is the speed of light, and β is the angle between spin and magnetic dipole axes; the value of β lies within 0–90 deg.

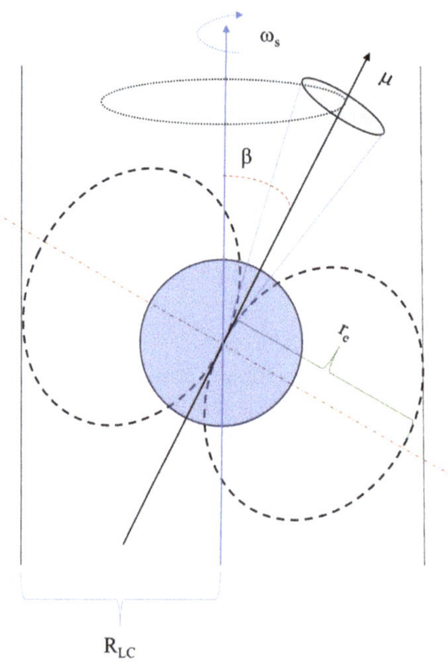

Figure 1. Scheme of rotating magnetized neutron star and its axes. In the figure denoted: μ is a magnetic dipole moment of NS, ω_{s} is an angular rotation velocity, β is an angle between spin and magnetic dipole axes, r_{e} is an equatorial radius of the magnetic field of NS, and R_{LC} is a radius of the light cylinder of NS.

The expression for rotational energy loss is the following:

$$\dot{E}_{\text{obs}} = I\omega_{\text{s}}\dot{\omega}_{\text{s}} = -I\frac{4\pi^2 \dot{P}_{\text{s}}}{P_{\text{s}}^3},\tag{2}$$

where I is a moment of inertia of NS, $\dot{\omega}_{\text{s}} = -2\pi\dot{P}_{\text{s}}/P_{\text{s}}^2$ is a derivative of the angular rotation velocity, $\dot{P}_{\text{s}}$ is a derivative of the spin period, i.e., rotational spin-up or spin-down.

Solving the system of Equations (1) and (2) by equating the losses, with μ_{ns} and using canonical values for NS (see Section 2.2), one can derive an expression for B_{ns} as

$$B_{\text{ns}}\sin\beta = 3.2 \times 10^{19}\sqrt{P_{\text{s}}\dot{P}_{\text{s}}} \quad \text{G}.\tag{3}$$

Equation (3) is a basic expression for estimating the magnetic field strength for rotation-powered pulsars. As seen from the equations above, $\dot{E}_{\text{md}}$ significantly depends on angle β, reaching the maximal energy loss $\dot{E}_{\text{md}}^{(\text{max})}$ with orthogonal axes ($\beta = 90°$). Indeed, in most cases, the magnetic field strength for radio pulsars is estimated by accepting the $\dot{E}_{\text{md}}^{(\text{max})}$ case; however, angle β can vary widely from the maximum value. Thus, it is crucial to correctly estimate the angle between spin and magnetic dipole axes to evaluate $B_{\text{ns}}(\beta)$ for rotation-powered pulsars.

Various methods for estimating the β-parameter have been previously proposed in the literature [8–11]. Here, we offer a relatively simple method based on a geometric approach for calculating the angle between the spin and magnetic dipole axes of a neutron star (NS) in the ejector stage. Section 2.1 outlines a basic concept and geometry for β. On its basis, in Section 2.2, we evaluate the surface magnetic field strength $B_{ns}(\beta)$ for a population of known neutron stars from the ATNF pulsar catalog. We provide the main results with a discussion in Sections 3 and 4.

The obtained data can help study properties and geometry of NS magnetic fields [12], study and model pulsar spin evolution [13], investigate stellar evolution in the late stages [14], etc.

2. Methods

2.1. Estimation of β-Parameter

According to [15], an equation of magnetic field lines, based on an assumption of a dipole magnetic field, in polar coordinates expressed as

$$r = r_e \left(\cos\phi\right)^2 , \tag{4}$$

where r_e is an equatorial radius of the magnetic field, corresponding to NS magnetosphere radius (see Figure 2), ϕ is an angle measured from the magnetic equator r_e towards the magnetic pole.

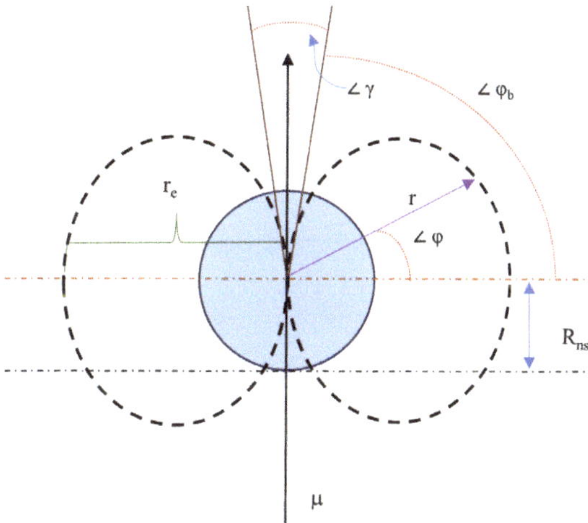

Figure 2. Scheme of a neutron star magnetosphere. In the figure denoted: R_{ns} is a radius of the neutron star, r is a radius vector of the magnetosphere, ϕ is an angle measured from the magnetic equator r_e towards the magnetic pole, with μ being an axis of a magnetic dipole moment of NS, γ is an opening angle of emission cone, ϕ_b corresponds to the angle between the magnetic equator and lateral surface of the emission cone, and r_e is an equatorial radius of the magnetic field of NS.

Using an Equation (4), it is possible to find an opening angle of emission cone γ assuming that the radio pulsar magnetosphere is limited by a light cylinder with radius R_{LC}, i.e., $r_e = R_{LC} = c \times P_s/2\pi$ [16]. At the base of the emission cone, the radius vector of the magnetic field line corresponds to the NS radius $r_b = R_{ns}$.

Solving Equation (4), we can find ϕ_b corresponding to the angle between the magnetic equator and lateral surface of the emission cone (see Figure 2).

$$\phi_b = \arccos\left[\left(\frac{R_{ns} \times 2\pi}{c \times P_s}\right)^{1/2}\right] . \tag{5}$$

Using Equation (5), one can find an opening angle of the emission cone γ:

$$\gamma = 2 \times \left(90° - \phi_b\right) . \tag{6}$$

At the next step, one can consider the dihedral angle (Figure 3) formed by two emission cone guides d and the rotation axis ω_s. Then, the linear diameter a of the emission cone at a distance d will be expressed as follows (via the cosine theorem):

$$\begin{cases} a^2 = 2d^2(1 - \cos\gamma) \\ a^2 = 2(d\sin\beta)^2 (1 - \cos\varepsilon) . \end{cases} \tag{7}$$

Solution of the Equation (7) with respect to parameter $\sin\beta$ gives

$$\sin\beta = \sqrt{\frac{1 - \cos\gamma}{1 - \cos\varepsilon}} . \tag{8}$$

The angle ε can be estimated by means of a pulse profile from observations of radio pulsars (see Figure 3 and Equation (9)). We suppose that a diameter a of the emission cone at a distance d covers the circle (dotted) around the larger cone formed by the magnetic axis μ rotating around the spin axis ω of the NS. The diameter of the cone a can be expressed in terms of the width of the observed pulse profile since the start and end times of the passage of the base of the emission cone through the observer correspond to the beginning (rise) and end (fall) of the curve in the pulse profile. We use data of the w_{10} parameter from the pulsar catalog, which is a pulse width at 10% of the peak of intensity [17] supposing that w_{10} approximately corresponds to the "travel" time (t) of the emission cone passing through an observer

$$\varepsilon = \omega_s t = \frac{2\pi}{P_s} t \simeq \frac{2\pi}{P_s} w_{10} . \tag{9}$$

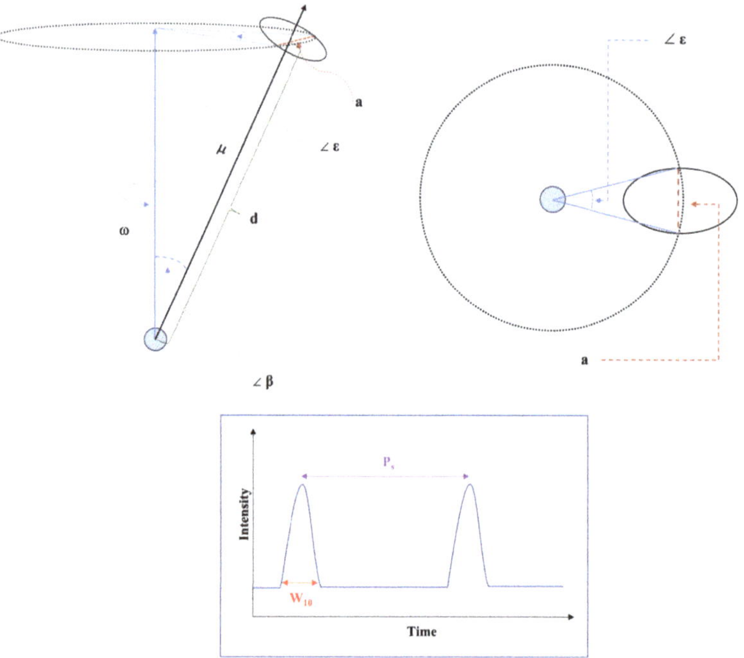

Figure 3. Scheme of the opening angle of the emission cone and the angle β between spin and magnetic dipole axes (**left** panel), and its view from the top (**right** panel). In the figure denoted: a is the linear diameter of the emission cone, and d is the distance. The rest of the symbols are identical to the ones in Figures 1 and 2. Bottom panel shows the pulse profile with period (P_s) and width of individual pulse at 10% of maximal intensity (w_{10}).

2.2. Data Selection and Evaluation of $B_{ns}(\beta)$

We use data from the ATNF pulsar catalog, available online[1] to apply our method to interested pulsars. The catalog contains information on rotation-powered pulsars and counts for over 3000 objects, and it is the most extensive database that provides information about radio pulsars. The catalog is maintained by the Australia Telescope National Facility (ATNF). The catalog includes detailed information about pulsars, such as their positions, rotation periods, spin-down rates, dispersion measures, and other relevant parameters. The ATNF pulsar catalog was initially compiled using data from the Parkes radio telescope in Australia. Over time, data from other radio telescopes in Australia and worldwide were incorporated into the catalog. The catalog is regularly updated as new observations are made, and new pulsars are discovered; see [2] for more details.

The data selection was carried out according to the following criteria:

- Objects with known spin period P_s;
- Exclusion of objects with $P_s > 2\,\mathrm{s}$;
- Known spin-down rate $\dot{P}_s$;
- Known w_{10} parameter.

The sample resulted in 1468 objects from the ATNF pulsar catalog, which are NS with $P_s < 2\,\mathrm{s}$ and the above parameters presented.

Some parameters of NS, such as radius R_{ns}, mass M_{ns}, and moment of inertia I are in a narrow range close to canonical values; their variations should influence the evaluation insignificantly [1,15]. Therefore, in our calculations, we use canonical values of mass, radius, and moment of inertia as $R_{ns} \simeq 10^6\,\mathrm{cm}$, $M_{ns} \simeq 1.4 M_\odot$, $I \simeq 10^{45}\,\mathrm{g\,cm^2}$, correspondingly.

3. Results

Table 1 (in full available in a machine-readable format) summarizes our calculations of β and $B_{ns}(\beta)$ parameters for the chosen population. The statistics on calculated β and $B_{ns}(\beta)$ is given in Tables 2 and 3.

Table 1. Sample of radio pulsars with estimated β and $B_{ns}(\beta)$. Other parameters (coordinates, P_s, $\dot{P}_s$, w_{10}) are extracted from ATNF pulsar catalog [17]. The full table containing data on 1468 objects is available online in a machine-readable format (Supplementary Table S1).

No.	Name PSR	RA J2000	DEC J2000	P_s (s)	$\dot{P}_s$ (s/s)	w_{10} (ms)	β (deg)	$B_{ns}(\beta)$ ($\times 10^{12}$ G)
1	J0006+1834	00:06:04.8	+18:34:59.0	0.69	2.10e-15	195.0	1.29	54.3
2	B0011+47	00:14:17.7	+47:46:33.4	1.24	5.64e-16	142.5	2.11	23
3	J0026+6320	00:26:50.5	+63:20:00.8	0.32	1.51e-16	48.0	3.22	3.94
4	B0031-07	00:34:08.8	−07:21:53.4	0.94	4.08e-16	120.0	2.20	16.4
5	J0038-2501	00:38:10.2	−25:01:30.7	0.26	7.60e-19	15.0	9.01	0.093
...	...	...	...	...	...	...	...	...

Table 2. Statistics for the β-parameter, an angle between spin and magnetic dipole axes, for different groups of radio pulsars.

Puls. Group	β Min	β Max	β Mean	β Median	σ	Total
$P_s < 0.01$	9.41	68.78	25.15	20.92	13.93	94
$0.01 < P_s < 0.1$	3.22	61.84	13.33	9.46	10.51	73
$0.1 < P_s < 2$	0.61	47.51	7.83	6.98	4.99	1301
All	0.61	68.78	9.21	7.43	7.65	1468

4. Discussion

In our paper, we used the classical dipole model of the radio pulsar magnetosphere proposed by [16]. In this case, the magnetosphere of a neutron star has a dipole structure co-rotating with a pulsar. It is limited by the so-called light cylinder on which the linear velocity of the magnetic field lines reaches the speed of light. This model is canonical and relevant to this day [18].

Indeed, the width of the pulse profile of radio pulsars can vary depending on the frequency (wavelength) of the observed flux. However, significant deviations in the profile width are observed at lower frequencies (<200 MHz). According to [19], this phenomenon is present because the light cone becomes wider when observed at lower frequencies, thereby seeing areas further from the pulsar's surface where the opening angle of the closed magnetic field lines is becoming broader. However, for higher frequencies (>200 MHz), this effect can be neglected [19]. The values of w_{10} in the ATNF Pulsar Catalog are the average profile width in the range of frequencies between 400–2000 MHz.

We showed that estimates of the surface magnetic field strength (B_{ns}) for a population of known neutron stars in the radio pulsar (ejector) stage should depend essentially on the angle β between spin and magnetic dipole axes of a neutron star. These estimates may differ by order of magnitude from those without considering the angle β (see Figure 4). The proposed method can be used when considering only rotation-powered pulsars with a MDR energy loss mechanism. This is not the case for the long-periodic pulsars with $P_s > 2$ s; therefore, we sharply cut off such objects in Figure 4, although borderline transition cases may occur individually.

Within the framework of the proposed technique, it is not possible to estimate the evolution of β over time since these changes are associated with changes in the flow of currents in the core of NS and the interaction of the magnetosphere with the surrounding plasma [20]. Nevertheless, we can compare our results against angles obtained within the framework of other methods.

As was mentioned in Section 1 there are several approaches for estimating the β-parameter. They can be conditionally divided into two groups: geometric and polarimetric methods. The first is based on different geometric models for NS magnetic field and emission cone. Our method also belongs to the first group. The second is based on measuring the position angle of linear polarization from radio pulsars, which depends on β [21]. Interest in comparing β from these two approaches resulted in the following consideration.

In recent articles [5,11,22] an estimation of β-parameter was obtained within geometric method based on spherical trigonometry. A polar cap model was used by authors with an assumption that the line-of-sight passes through the center of the emission cone. The comparison between our data and data from [5,11,22] is shown in Figure 6 for matched 1242 and 246 radio pulsars and their statistics are given in Table 4. In most cases, the difference in estimates ($\Delta\beta$ median) does not exceed 5 deg and is mainly caused by the difference in the methods (models) used. Negative and positive trends can be noticed correspondingly between our data and data by Ken'ko et al. 2023 [5] (Figure 6, left panel), with a vertical dotted line approximately marking the spin period where two methods give similar β estimation. Again, this is due to differences in geometric approaches for β estimation since the data themselves for both methods were taken from the same catalog (ATNF Pulsar Catalogue). No trends are seen between our data and the data by Nikitina et al. 2017 [22] (Figure 6, right panel), where blue dots are systematically positioned above red ones for matches pulsars. While these authors use analogous methods based on spherical trigonometry, their sample is relatively small, so the trends may not have enough data to manifest. Another reason could be in the data themselves, since in [22] the authors have used data from their observational facilities (Pushchino Radio Astronomy Observatory).

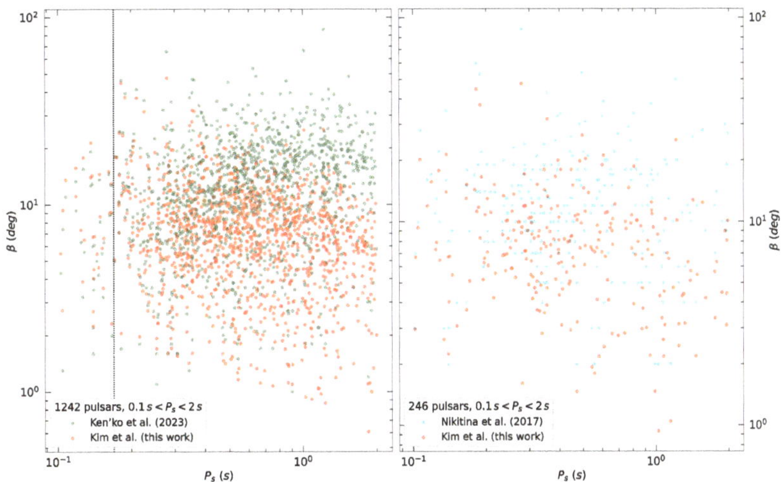

Figure 6. A comparison of β-parameter estimation for a radio pulsar population obtained by Ken'ko et al. (2023) [5] (green dots, 1242 objects, **left** panel) and Nikitina et al. (2017) [22] (light blue dots, 246 objects, **right** panel), and by our method (red dots, both panels). There are negative and positive trends between our data and data by Ken'ko et al. (2023) (**left** panel) correspondingly, with a vertical dotted line approximately marking the spin period where two methods give similar β estimations. No trends are seen between our data and the data by Nikitina et al. (2017) (**right** panel).

Table 4. Statistics for the comparison of β obtained from geometric approaches from Ken'ko et al. (2023) [5] and Nikitina et al. (2017) [22] with our data. All pulsars have spin periods between 0.1 s and 2 s.

Pulsar Sample	Selection Criteria	Pulsar Number	$\|\Delta\beta\|$ Min	$\|\Delta\beta\|$ Max	$\|\Delta\beta\|$ Mean	$\dfrac{\|\Delta\beta\|}{\text{Median}}$	σ
Ken'ko et al. (2023) [5]	P_s, w_{10}	1242	0.007338	61.145636	5.825957	4.61572	5.013604
Nikitina et al. (2017) [22]	$P_s, \dot{P}_s, w_{10}$	246	0.037164	40.488991	5.544564	4.094941	5.092834

We further attempted to compare estimates obtained by polarimetric studies to determine the angle β performed for only a small part of the radio pulsar population (see Table 5). This method is based on measuring the position angle of linear polarization and is more reliable than geometric approaches. However, for some objects, when observed in different wavelength ranges (frequencies), it can give a significant scatter, especially for larger β. For example, as shown in [22] for PSR B1055-52 (aka J1057-5226) β-parameter estimation at 10 cm wavelength gives $\beta_{10-\text{cm}} = 15$ deg, but estimation at 20 cm wavelength gives $\beta_{20-\text{cm}} = 24$ deg; for PSR B1702-19 (aka J1705-1906) $\beta_{10-\text{cm}} = 49$ deg, $\beta_{20-\text{cm}} = 70$ deg, etc. The larger scatter in $|\beta_{\text{pol}} - \beta_{\text{geom}}|$ between geometric and polarimetric methods is mainly due to the assumption that the line-of-sight passes through the center of the base of the emission cone. Thus, the geometric estimates are the lower limits for the measured angle β [5].

As also seen in Figure 4, we obtained 110 objects (7.5% from 1468 pulsar sample) with estimated magnetic fields exceeding the so-called quantum critical threshold $\sim$4.4 $\times 10^{13}$ G [23]. These are blue dots over the solid red line, and all (except one short-periodic source PSR B0540-69) belong to the population of middle-periodic pulsars. The maximal value $B_{\text{ns}}(\beta) \simeq 7.56 \times 10^{14}$ G refers to the pulsar PSR J1119-6127. According to [24] this radio pulsar demonstrates episodic SGR-like high-energy bursts reaching 2.8×10^{39} erg s^{-1} within 15–150 keV range. The magnetic field of the NS derived from analysis of PSR J1119-6127 during its burst activity corresponds to $B_{\text{ns}} \sim 10^{14}$ G [24] that agrees with our estimate within an order of magnitude. The analysis of the rest of the high-B sub-sample can be interesting from the point of a possible relation between high-B radio pulsars and the population of isolated X-ray pulsars [25]: anomalous X-ray pulsars (AXP), soft gamma-repeaters (SGR), etc.

Table 5. Comparison of the obtained data from this work (β_{geom}) with data obtained by polarization method (β_{pol}) from Nikitina et al. (2017) [22].

No.	Name PSR	β_{geom} (deg)	β_{pol} (deg)	$|\beta_{pol} - \beta_{geom}|$ (deg)
1	J0108-1431	3.93	11	7.07
2	B0656+14	5.18	17	11.82
3	J0905-5127	14.30	22	7.70
4	J1015-5719	2.25	5	2.75
5	B1055-52	6.62	15	8.38
6	J1349-6130	9.05	64	54.95
7	J1355-5925	6.49	10	3.51
8	B1509-58	3.07	10	6.93
9	B1702-19	10.41	49	38.59
10	J1702-4310	5.73	11	5.27
11	J1723-3659	17.96	28	10.04
12	B1800-21	2.64	12	9.36
13	B1822-14	5.07	8	2.93

As seen in Figure 5 and Table 1, with increasing spin period P_s, there is a tendency of the angle β to decrease. This agrees with the current view of the spin evolution of NS [20]: older neutron stars have lengthier spin periods and smaller values of β, excepting a millisecond pulsar population. According to [20] on the timescales 10^6–10^7 yr in the ejector stage a NS should align its magnetic and spin axes, i.e., the angle β tends to zero.

For the population of millisecond pulsars (MSPs), the evolution of the β-parameter may differ significantly from other radio-pulsar populations. The millisecond pulsars are neutron stars in close binary systems or descendants of close binary systems in the case of isolated MSPs, with a low-mass companion, where accretion flow from a normal companion recycled a NS to ultra-short spin periods [26]. Thus, MSPs are old neutron stars whose rotational evolution has gone all possible stages (*ejector* → *propeller* → *accretor*) and then came back to the *ejector* stage through accretion recycling [26].

According to [27], the initial ejector stage for a neutron star in a binary system (with a normal star companion) lasts 10^5–10^6 yr, that is much shorter than in the case of an isolated NS and order of magnitude shorter than the timescale needed for aligning magnetic and spin axes of NS in ejector stage (see previous paragraph). Therefore, a NS in a binary system can move on to the following evolutionary stages (*propeller* and *accretor*) from the ejector stage with a β-parameter, which is significantly different from a zero value. Moreover, according to [28], the magnetic and spin axes of a neutron star in the stage of accretion tend to an orthogonal position, i.e., β-parameter increases to 90 deg on the timescale $\sim 10^5$ yr. The maximal possible lifetime of a NS on the accretor stage in a low-mass binary system is comparable to the lifetime of its normal companion, $\sim(0.1$–$10) \times 10^9$ yr [29]. It exceeds the orthogonalization timescale by several orders of magnitude, sufficient to increase the β-parameter significantly. Thus, MSPs are old neutron stars that demonstrate large values of β-parameter compared to other types of radio pulsars in the ejector stage.

Supplementary Materials: The following supporting information can be downloaded at: https://www.mdpi.com/article/10.3390/universe9070334/s1, Table S1: Sample of radio pulsars with estimated β and $B_{ns}(\beta)$. Other parameters (coordinates, P_s, $\dot{P}_s$, w_{10}) are extracted from ATNF pulsar catalog [17].

Author Contributions: Conceptualization, writing—review and editing, project administration, V.K. and Y.A.; methodology, investigation, writing—original draft preparation, supervision, V.K.; validation, data curation, visualization, A.U.; formal analysis, resources, funding acquisition, Y.A. All authors have read and agreed to the published version of the manuscript.

Funding: This research has been funded by the Science Committee of the Ministry of Education and Science of the Republic of Kazakhstan (Grant No. AP09258811).

Acknowledgments: Authors acknowledge Remo Ruffini and the organizers of the conference "Prof. Remo Ruffini Festschrift" for the invitation to give a talk and the support provided during the online meeting. We thank the Editors and Referees for their comments and suggestions to improve the presentation of the results.

Conflicts of Interest: The authors declare no conflict of interest. The funders had no role in the design of the study; in the collection, analyses, or interpretation of data; in the writing of the manuscript; or in the decision to publish the results.

Abbreviations

The following abbreviations are used in this manuscript:

ATNF	Australia Telescope National Facility
AXP	Anomalous X-ray pulsar
deg	Degree (unit)
G	Gauss (unit)
MDR	Magnetic Dipole Radiation
ms	Millisecond (unit)
MSP	Millisecond pulsar
NS	Neutron star
SGR	Soft gamma-repeater
s	Second (unit)
s/s	Seconds per second (unit)
yr	Year (unit)

Note

[1] https://www.atnf.csiro.au/people/pulsar/psrcat/ (accessed on 2 June 2023).

References

1. Lyne, A.; Graham-Smith, F. *Pulsar Astronomy*; Cambridge University Press: Cambridge, UK, 2012.
2. Manchester, R.N.; Hobbs, G.B.; Teoh, A.; Hobbs, M. The Australia Telescope National Facility Pulsar Catalogue. *Astron. J.* **2005**, *129*, 1993–2006. [CrossRef]
3. Tan, C.M.; Bassa, C.G.; Cooper, S.; Dijkema, T.J.; Esposito, P.; Hessels, J.W.T.; Kondratiev, V.I.; Kramer, M.; Michilli, D.; Sanidas, S.; et al. LOFAR Discovery of a 23.5 s Radio Pulsar. *Astrophys. J.* **2018**, *866*, 54. [CrossRef]
4. Malov, I.F.; Marozava, H.P. On Braking Mechanisms in Radio Pulsars. *Astron. Rep.* **2022**, *66*, 25–31. [CrossRef]
5. Ken'ko, Z.V.; Malov, I.F. Evolution of the angles between magnetic moments and rotation axes in radio pulsars. *Mon. Not. R. Astron. Soc.* **2023**, *522*, 1826–1842. [CrossRef]
6. Gunn, J.E.; Ostriker, J.P. Magnetic Dipole Radiation from Pulsars. *Nature* **1969**, *221*, 454–456. [CrossRef]
7. Pacini, F. The Nature of Pulsar Radiation. *Nature* **1970**, *226*, 622–624. [CrossRef] [PubMed]
8. Kuz'min, D.K.; Dagkesamanskaya, I.M. Evaluation of the angles of magnetic axis relative to rotation axis of the Pulsars. *Astron. Lett.* **1983**, *9*, 149–154.
9. Lyne, A.G.; Manchester, R.N. The shape of pulsar radio beams. *Mon. Not. R. Astron. Soc.* **1988**, *234*, 477–508. [CrossRef]
10. Malov, I.F. Angle Between the Magnetic Field and the Rotation Axis in Pulsars. *Sov. Astron.* **1990**, *34*, 189.
11. Ken'ko, Z.V.; Malov, I.F. Comparison of the Angles between the Magnetic Moment and Rotation Axis for Two Groups of Radio Pulsars. *Astron. Rep.* **2022**, *66*, 669–692. [CrossRef]
12. Igoshev, A.P.; Popov, S.B.; Hollerbach, R. Evolution of Neutron Star Magnetic Fields. *Universe* **2021**, *7*, 351. [CrossRef]
13. Xie, Y.; Zhang, S.N. Rotational behaviors and magnetic field evolution of radio pulsars. *Astron. Nachrichten* **2014**, *335*, 775, [CrossRef]
14. Portegies Zwart, S.F.; Van den Heuvel, E.P.J. The origin of single radio pulsars. *New Astron.* **1999**, *4*, 355–363. [CrossRef]
15. Lipunov, V.M. *Astrophysics of Neutron Stars*; Springer: Berlin/Heidelberg, Germany; New York, NY, USA, 1992.
16. Goldreich, P.; Julian, W.H. Pulsar Electrodynamics. *Astrophys. J.* **1969**, *157*, 869. [CrossRef]

17. Manchester, R.N.; Hobbs, G.B.; Teoh, A.; Hobbs, M. VizieR Online Data Catalog: ATNF Pulsar Catalogue (Manchester+, 2005); VizieR Online Data Catalog: 2016; p. B/psr. Available online: https://ui.adsabs.harvard.edu/abs/2016yCat....102034M/abstract (accessed on 1 June 2023).
18. Rahaman, S.M.; Mitra, D.; Melikidze, G.I.; Lakoba, T. Pulsar radio emission mechanism—II. On the origin of relativistic Langmuir solitons in pulsar plasma. *Mon. Not. R. Astron. Soc.* **2022**, *516*, 3715–3727. [CrossRef]
19. Pilia, M.; Hessels, J.W.T.; Stappers, B.W.; Kondratiev, V.I.; Kramer, M.; van Leeuwen, J.; Weltevrede, P.; Lyne, A.G.; Zagkouris, K.; Hassall, T.E.; et al. Wide-band, low-frequency pulse profiles of 100 radio pulsars with LOFAR. *Astron. Astrophys.* **2016**, *586*, A92. [CrossRef]
20. Philippov, A.; Tchekhovskoy, A.; Li, J.G. Time evolution of pulsar obliquity angle from 3D simulations of magnetospheres. *Mon. Not. R. Astron. Soc.* **2014**, *441*, 1879–1887. [CrossRef]
21. Radhakrishnan, V.; Cooke, D.J. Magnetic Poles and the Polarization Structure of Pulsar Radiation. *Astrophys. Lett.* **1969**, *3*, 225.
22. Nikitina, E.B.; Malov, I.F. Magnetic fields of radio pulsars. *Astron. Rep.* **2017**, *61*, 591–611. [CrossRef]
23. Peng, Q.H.; Tong, H. The physics of strong magnetic fields in neutron stars. *Mon. Not. R. Astron. Soc.* **2007**, *378*, 159–162. [CrossRef]
24. Göğüş, E.; Lin, L.; Kaneko, Y.; Kouveliotou, C.; Watts, A.L.; Chakraborty, M.; Alpar, M.A.; Huppenkothen, D.; Roberts, O.J.; Younes, G.; et al. Magnetar-like X-Ray Bursts from a Rotation-powered Pulsar, PSR J1119-6127. *Astrophys. J. Lett.* **2016**, *829*, L25. [CrossRef]
25. Benli, O.; Ertan, Ü. On the evolution of high-B radio pulsars with measured braking indices. *Mon. Not. R. Astron. Soc.* **2017**, *471*, 2553–2557. [CrossRef]
26. Lorimer, D.R. Binary and Millisecond Pulsars. *Living Rev. Relativ.* **2008**, *11*, 8. [CrossRef] [PubMed]
27. Urpin, V.; Geppert, U.; Konenkov, D. Magnetic and spin evolution of neutron stars in close binaries. *Mon. Not. R. Astron. Soc.* **1998**, *295*, 907–920. [CrossRef]
28. Biryukov, A.; Abolmasov, P. Magnetic angle evolution in accreting neutron stars. *Mon. Not. R. Astron. Soc.* **2021**, *505*, 1775–1786. [CrossRef]
29. Kiziltan, B.; Thorsett, S.E. Millisecond Pulsar Ages: Implications of Binary Evolution and a Maximum Spin Limit. *Astrophys. J.* **2010**, *715*, 335–341. [CrossRef]

MDPI

Article

Neutron Star Binaries Produced by Binary-Driven Hypernovae, Their Mergers, and the Link between Long and Short GRBs

Laura M. Becerra [1,2,*,†], Chris Fryer [3,†], Jose F. Rodriguez [1,2,*,†], Jorge A. Rueda [2,4,5,6,7*,†] and Remo. Ruffini [2,4,8,†]

1   GIRG, Escuela de Física, Universidad Industrial de Santander, Bucaramanga 680002, Colombia
2   ICRANet, Piazza della Repubblica 10, 65122 Pescara, Italy
3   CCS-2, Los Alamos National Laboratory, Los Alamos, NM 87545, USA
4   ICRA, Dip. di Fisica, Sapienza Università di Roma, P.le Aldo Moro 5, 00185 Rome, Italy
5   ICRANet-Ferrara, Dip. di Fisica e Scienze Della Terra, Università Degli Studi di Ferrara, Via Saragat 1, 44122 Ferrara, Italy
6   Dip. di Fisica e Scienze della Terra, Università degli Studi di Ferrara, Via Saragat 1, 44122 Ferrara, Italy
7   INAF, Istituto de Astrofisica e Planetologia Spaziali, Via Fosso del Cavaliere 100, 00133 Rome, Italy
8   INAF, Viale del Parco Mellini 84, 00136 Rome, Italy
*   Correspondence: laura.becerra7@correo.uis.edu.co (L.M.B.); joferoru@gmail.com (J.F.R.); jorge.rueda@icra.it (J.A.R.)
†   These authors contributed equally to this work.

Abstract: The binary-driven hypernova (BdHN) model explains long gamma-ray bursts (GRBs) associated with supernovae (SNe) Ic through physical episodes that occur in a binary composed of a carbon-oxygen (CO) star and a neutron star (NS) companion in close orbit. The CO core collapse triggers the cataclysmic event, originating the SN and a newborn NS (hereafter νNS) at its center. The νNS and the NS accrete SN matter. BdHNe are classified based on the NS companion fate and the GRB energetics, mainly determined by the orbital period. In BdHNe I, the orbital period is of a few minutes, so the accretion causes the NS to collapse into a Kerr black hole (BH), explaining GRBs of energies $>10^{52}$ erg. BdHN II, with longer periods of tens of minutes, yields a more massive but stable NS, accounting for GRBs of 10^{50}–10^{52} erg. BdHNe III have still longer orbital periods (e.g., hours), so the NS companion has a negligible role, which explains GRBs with a lower energy release of $<10^{50}$ erg. BdHN I and II might remain bound after the SN, so they could form NS-BH and binary NS (BNS), respectively. In BdHN III, the SN likely disrupts the system. We perform numerical simulations of BdHN II to compute the characteristic parameters of the BNS left by them, their mergers, and the associated short GRBs. We obtain the mass of the central remnant, whether it is likely to be a massive NS or a BH, the conditions for disk formation and its mass, and the event's energy release. The role of the NS nuclear equation of state is outlined.

Keywords: neutron stars; gamma-ray burst; close binaries

Citation: Becerra, L.M.; Fryer, C.; Rodriguez, J.F.; Rueda, J.A.; Ruffini, R. Neutron Star Binaries Produced by Binary-Driven Hypernovae, Their Mergers, and the Link between Long and Short GRBs. *Universe* **2023**, *9*, 332. https://doi.org/10.3390/universe9070332

Academic Editor: Fridolin Weber

Received: 15 May 2023
Revised: 8 July 2023
Accepted: 9 July 2023
Published: 12 July 2023

1. Introduction

Gamma-ray bursts (GRBs) are classified using the time (in the observer's frame) T_{90}, in which 90% of the observed isotropic energy (E_{iso}) in the gamma-rays is released. Long GRBs have $T_{90} > 2$ s and short GRBs, $T_{90} < 2$ s [1–5]. The two types of sources, short and long GRBs, are thought to be related to phenomena occurring in gravitationally collapsed objects, e.g., stellar-mass black holes (BHs) and neutron stars (NSs).

For short GRBs, mergers of binary NSs (BNSs) and/or NS-BH were soon proposed as progenitors [6–9]). For long bursts, the core-collapse of a single massive star leading to a BH (or a magnetar), a *collapsar* [10], surrounded by a massive accretion disk has been the traditional progenitor (see, e.g., [11,12], for reviews). The alternative binary-driven hypernova (BdHN) model exploits the increasing evidence for the relevance of a binary progenitor for long GRBs, e.g., their association with Ic-type supernovae (SNe) [13–16], proposing a binary system composed of a carbon-oxygen star (CO) and an NS companion for long GRBs. We refer the reader to [17–23] for theoretical details on the model.

In this article, we are interested in the direct relationship between long and short GRBs predicted by the BdHN scenario. The CO undergoes core collapse, ejecting matter in a supernova (SN) explosion and forming a newborn NS (νNS) at its center. The NS companion attracts part of the ejected material leading to an accretion process with high infalling rates. Also, the νNS gains mass via a fallback accretion process. The orbital period is the most relevant parameter for the CO-NS system's fate. In BdHN of type I, the NS reaches the critical mass, gravitationally collapsing into a Kerr BH. It occurs for short orbital periods (usually a few minutes) and explains GRBs with energies above 10^{52} erg. In BdHN II, the orbital period is larger, up to a few tens of minutes, so the accretion rate decreases, and the NS becomes more massive but remains stable. These systems explain GRBs with energies 10^{50}–10^{52} erg. In BdHN III, the orbital separation is still larger; the NS companion does not play any role, and the energy release is lower than 10^{50} erg. If the binary is not disrupted by the mass loss in the SN explosion (see [20] for details), a BdHN I produces a BH-NS, whereas a BdHN II produces a BNS. In BdHN III, the SN is expected to disrupt the system. Therefore, in due time, the mergers of NS-BHs left by BdHNe I and of BNS left by BdHNe II are expected to lead to short GRBs.

Short GRBs from BNS mergers have been classified into short gamma-ray flashes (S-GRFs) and authentic short GRBs (S-GRBs), depending on whether the central remnant is an NS or a BH, respectively [24]. Two different subclasses of short GRBs from BNS mergers have been electromagnetically proposed [20,24,25]:

(1) *Authentic short GRBs (S-GRBs)*: short bursts with isotropic energy $E_{\mathrm{iso}} \gtrsim 10^{52}$ erg and peak energy $E_{p,i} \gtrsim 2$ MeV. They occur when a BH is formed in the merger, which is revealed by the onset of a GeV emission (see [25–27]). Their electromagnetically inferred isotropic occurrence rate is $\rho_{\mathrm{S-GRB}} \approx \left(1.9^{+1.8}_{-1.1}\right) \times 10^{-3}$ Gpc^{-3} year^{-1} [24]. The distinct signature of the formation of the BH, namely the observation of the 0.1–100 GeV emission by the *Fermi*-LAT, needs the presence of baryonic matter interacting with the newly-formed BH, e.g., via an accretion process (see, e.g., [26,28]).

(2) *Short gamma-ray flashes (S-GRFs)*: short bursts with $E_{\mathrm{iso}} \lesssim 10^{52}$ erg and $E_{p,i} \lesssim 2$ MeV. They occur when no BH is formed in the merger, i.e., when it leads to a massive NS. Their U-GRB electromagnetically inferred isotropic occurrence rate is $\rho_{\mathrm{S-GRF}} \approx 3.6^{+1.4}_{-1.0}$ Gpc^{-3} year^{-1} [24].

(3) *Ultrashort gamma-ray flashes (U-GRFs)*: in [20], it has been advanced a new class short bursts, the ultrashort GRBs (U-GRBs) produced by NS-BH binaries when the merger leaves the central BH with very little or completely without surrounding matter. An analogous system could be produced in BNS mergers. We shall call these systems ultrashort GRFs, for short U-GRFs. Their gamma-ray emission is expected to occur in a prompt short radiation phase. The post-merger radiation is drastically reduced, given the absence of baryonic matter to power an extended emission. A *kilonova* can still be observed days after the merger, in the infrared, optical, and ultraviolet wavelengths, produced by the radioactive decay of r-process yields [29–32]. Kilonova models used a *dynamical* ejecta composed of matter expelled by tides prior or during the merger, and a *disk-wind* ejecta by matter expelled from post-merger outflows in accretion disks [33], so U-GRFs are expected to have only the dynamical ejecta kilonova emission.

We focus on the BNSs left by BdHNe II and discuss how their properties impact the subsequent merger process and the associated short GRB emission, including their GW radiation. Since an accretion disk around the central remnant of a BNS merger, i.e., a newborn NS or a BH, is an important ingredient in models of short GRBs (see, e.g., [34] and references therein), we give some emphasis to the conditions and consequences for the merger leaving a disk. We study BNSs formed through binary evolution channels. Specifically, we expect these systems to form following a binary evolution channel similar to that of two massive stars leading to stripped-envelope binaries, described in previous studies (e.g., [35,36]). In this process, the CO star undergoes mass loss in multiple mass-transfer and common-envelope phases through interactions with the NS companion (see, e.g., [37–39]). This leads to removing the H/He layers of the secondary star, which ends up as a CO star. Recently, significant progress has been made in the study of alternative evolution channels for the progenitor of BNSs, such as hierarchical systems involving triple and quadrupole configurations [40,41], which are motivated by the presence of massive stars in multiple systems [42]. These systems are out of the scope of this study.

The article is organized as follows. In Section 2, we discuss the numerical simulations of BdHNe and specialize in an example of a BNS led by a BdHN II. Section 3 introduces a theoretical framework to analyze the BNS merger outcome configuration properties based on the conservation laws of baryon number, angular momentum, and mass-energy. We present in Section 4 a specific example analyzing a BNS merger using the above-mentioned theoretical framework, including estimates of the energy and angular momentum release. We include the radiation in gravitational waves (GWs) and estimate its detection by current facilities. Section 5 presents a summary and the conclusions of this work.

2. A BNS Left by a BdHN II

Figure 1 shows a snapshot of the mass density with the vector velocity field at the binary's equatorial plane some minutes after the CO collapse and the expansion of the SN ejecta. The system's evolution was simulated with an SPH code, where the NS companion and the νNS are point particles that interact gravitationally with the SPH particles of the SN ejecta. For details of these numerical simulations, we refer to [23,43]. In these simulations, the influence of the star's magnetic field has be disregarded, as the magnetic pressure remains significantly lower than the random pressure exerted on the infalling material. The simulation of Figure 1 corresponds to a CO-NS for a CO star evolved from a zero-age main-sequence (ZAMS) star of $M_{\text{zams}} = 15\ M_\odot$. The CO mass is about 3.06 $M_\odot$, whose core collapse leaves a 1.4 $M_\odot$ νNS and ejects 1.66 $M_\odot$. The NS companion's initial mass is 1.4 $M_\odot$, and the initial binary period of the system is about 4.5 min.

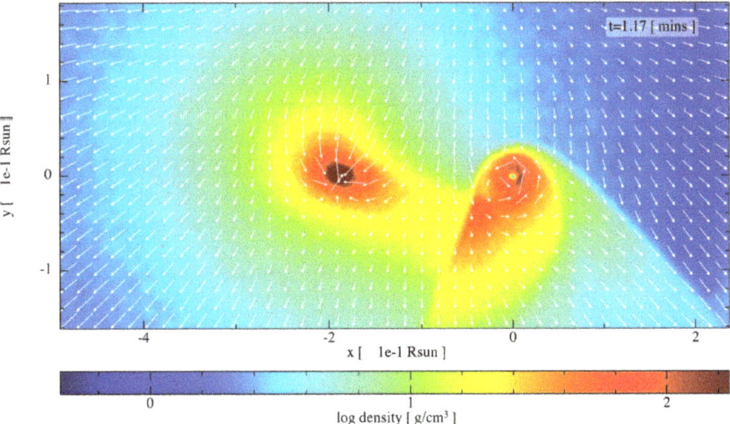

Figure 1. Massdensity snapshots and velocity field on the orbital plane of a BdHN for a CO left by a $M_{\text{zams}} = 15\ M_\odot$ and a 1.4 $M_\odot$ NS companion, with an initial orbital period of about 4.5 min. We follow the expansion of the SN ejecta in the presence of the NS companion and the $\nu-$NS with a smoothed particle hydrodynamic (SPH) code. It is clear that a disk with opposite spins has formed around both stars.

From the accretion rate on the NSs, we have calculated the evolution of the mass and angular momentum of the binary components (see [43], for details). Table 1 summarizes the final parameters of the νNS and the NS, including the gravitational mass, m, dimensionless angular momentum, j, angular velocity, Ω, equatorial radius, R_{eq}, and moment of inertia,

1. These structure parameters have been calculated with the RNS code [44] and using the GM1 [45,46] and TM1 [47] EOS (see Table 2 for details of the EOS). The BNS left by the BdHN II event has a period $P_{orb} = 14.97$ min, orbital separation $a_{orb} \approx 2 \times 10^{10}$ cm, and eccentricity $e = 0.45$.

Table 1. BNS produced by a BdHN II originated in a CO-NS with an orbital period of 4.5 min. The CO star mass is 3.06 $M_\odot$, obtained from the stellar evolution of a ZAMS star of $M_{zams} = 15\ M_\odot$, and the NS companion has 1.4 $M_\odot$. The numerical smoothed-particle hydrodynamic (SPH) simulation follows the SN produced by the CO core collapse and estimates the accretion rate onto the νNS and the NS companion. The structure parameters of the NSs are calculated for the GM1 and TM1 EOS. We refer to [43] for additional details.

	m [$M_\odot$]	j	Ω [s^{-1}]	R_{eq} [km]	I [g cm^2]	Ω [s^{-1}]	R_{eq} [km]	I [g cm^2]
				GM1 EOS			**TM1 EOS**	
νNS	1.505	0.259	1114.6	14.03	2.04×10^{45}	1077.1	14.47	2.11×10^{45}
NS	1.404	-0.011	-52.14	14.01	1.85×10^{45}	-56.6	14.49	1.93×10^{45}

Table 2. Properties of the selected EOS. From left to right: maximum stable mass of non-rotating configurations, uniformly rotating configurations, set by the maximum mass of the Keplerian/mass-shedding sequence and the corresponding angular velocity.

EOS	$M_{max}^{j=0}$ [$M_\odot$]	$M_{max}^{j_{kep}}$ [$M_\odot$]	Ω_{kep}^{max} [s^{-1}]
GM1	2.38	2.84	1.001×10^4
TM1	2.19	2.62	8.83×10^3

3. Inferences from Conservation Laws

We analyze the properties of the central remnant NS formed after the merger. We use the conservation laws of baryon number, energy, and angular momentum for this aim.

3.1. Baryon Number Conservation

The total baryonic mass of the system must be conserved, so the binary baryonic mass, M_b, will redistribute among that of the postmerger's central remnant, $m_{b,c}$; the ejecta's mass, m_{ej}, which is unbound to the system; and the matter kept bound to the system, e.g., in the form of a disk of mass m_d. Therefore, we have the constraint

$$M_b = m_{b,c} + m_{ej} + m_d, \quad M_b = m_{b,1} + m_{b,2}. \tag{1}$$

For a uniformly rotating NS, the relation among its baryonic mass, $m_{b,i}$, gravitational mass, m_i, and angular momentum J_i, is well represented by the simple function

$$\frac{m_{b,i}}{M_\odot} \approx \frac{m_i}{M_\odot} + \frac{13}{200} \left(\frac{m_i}{M_\odot} \right)^2 \left(1 - \frac{1}{130} j_i^{1.7} \right), \quad i = 1, 2, c, \tag{2}$$

where $j_i \equiv cJ_i/(GM_\odot^2)$, which fits numerical integration solutions of the axisymmetric Einstein equations for various nuclear EOS, with a maximum error of 2% [48]. Thus, Equation (2) is a nearly universal, i.e., EOS-independent, formula. Equation (2) applies to the merging components ($i = 1, 2$) as well as to the central remnant ($i = c$).

3.2. Angular Momentum Conservation

We can make more inferences about the merger's fate from the conservation of angular momentum. The angular momentum of the binary during the inspiral phase is given by

$$J = \mu r^2 \Omega + J_1 + J_2, \quad J_i = \frac{2}{5}\kappa_i m_i R_i^2 \Omega_i, \quad i = 1,2, \tag{3}$$

where r is the orbital separation, $\mu = m_1 m_2 / M$ is the reduced mass, $M = m_1 + m_2$ is the total binary mass, and $\Omega = \sqrt{GM/r^3}$ is the orbital angular velocity. The gravitational mass and stellar radius of the i-th stellar component are, respectively, m_i and R_i; J_i is its angular momentum, Ω_i its angular velocity, and κ_i is the ratio between its moment of inertia to that of a homogeneous sphere. We adopt the convention $m_2 \leq m_1$. After the merger, the angular momentum is given by the sum of the angular momentum of the central remnant, the disk, and the ejecta. Angular momenta conservation implies that the angular momenta at merger, J_{merger}, equals that of the final configuration plus losses:

$$J_{\text{merger}} = J_c + J_d + \Delta J, \tag{4}$$

where J_c and J_d are, respectively, the angular momenta of the central remnant and the eventual surrounding disk, ΔJ accounts for angular momentum losses, e.g., via gravitational waves, and we have neglected the angular momentum carried out by the ejecta since it is expected to have small mass $\sim 10^{-4}$–10^{-2} $M_\odot$. Simulations suggest that this ejecta comes from interface of the merger, where matter is squeezed and ejected perpendicular to the orbital plane, see, e.g., [49,50]. The definition of the merger point will be discussed below.

The angular momentum of the binary at the merger point is larger than the maximum value a uniformly rotating NS can attain, i.e., the angular momentum at the Keplerian/mass-shedding limit, J_K. Thus, the remnant NS should evolve first through a short-lived phase that radiates the extra angular momentum over that limit and enters the rigidly rotating stability phase from the mass-shedding limit. Thus, we assume the remnant NS after that transition phase starts its evolution with angular momentum

$$J_c = J_K \approx 0.7 \frac{Gm_c^2}{c}. \tag{5}$$

Equation (5) fits the angular momentum of the Keplerian sequence from full numerical integration of the Einstein equations and is nearly independent of the nuclear EOS (see, e.g., [48] and references therein). Therefore, the initial dimensionless angular momentum of the central remnant is

$$j_c = \frac{cJ_c}{GM_\odot^2} \approx 0.7 \left(\frac{m_c}{M_\odot} \right)^2. \tag{6}$$

We model the disk's angular momentum as a ring at the remnant's inner-most stable circular orbit (ISCO). Thus, we use the formula derived in Cipolletta et al. [51], which fits, with a maximum error of 0.3%, the numerical results of the angular momentum per unit mass of a test particle circular orbit in the general relativistic axisymmetric field of a rotating NS. Within this assumption, the disk's angular momentum is given by

$$J_d = J_{\text{ISCO}} \approx \frac{G}{c} m_c m_d \left[2\sqrt{3} - 0.37 \left(\frac{j_c}{m_c/M_\odot} \right)^{0.85} \right]. \tag{7}$$

Notice that Equation (7) reduces to the known result for the Schwarzschild metric for vanishing angular momentum, as it must. However, it differs from the result for the Kerr metric, which tells us that the Kerr metric does not describe the exterior spacetime of a rotating NS (see [51] for a detailed discussion).

The estimate of J_{merger} requires the knowledge of the merger point, which depends on whether or not the binary secondary becomes noticeably deformed by the tidal forces.

When the binary mass ratio $q \equiv m_2/m_1$ is close or equal to 1, the stars are only deformed before the point of contact [52]. Therefore, for $q \approx 1$, we can assume the point of the merger as the point of contact

$$r_{\text{merger}} \approx r_{\text{cont}} = \frac{(\mathcal{C}_2 + q\mathcal{C}_1)}{(1+q)\mathcal{C}_1\mathcal{C}_2} \frac{GM}{c^2}, \tag{8}$$

where $\mathcal{C}_{1,2} \equiv Gm_{1,2}/(c^2 R_{1,2})$ is the compactness of the BNS components.

When the masses are different, if we model the stars as Newtonian incompressible spheroids, there is a minimal orbital separation r_{ms}, below which no equilibrium configuration is attainable, i.e., one star begins to shed mass to the companion due to the tidal forces. In this approximation, $r_{\text{ms}} \approx 2.2 q^{-1/3} R_2$ [53]. Numerical relativity simulations of BH-NS quasi-equilibrium states suggest that the mass-shedding occurs at a distance (see [54] and references therein) of

$$r_{\text{ms}} \approx (0.270)^{-2/3} q^{-1/3} R_2. \tag{9}$$

Our analysis adopts the mass-shedding distance of Equation (9). For a system with $q = 0.7$ (similar mass ratio of the one in Table 1), we have found that the less-compact star begins to shed mass before the point of contact, independently of the EOS, which agrees with numerical relativity simulations. Consequently, for non-symmetric binaries $q < 1$, we define the merging at the point as the onset of mass-shedding, $r_{\text{merger}} \approx r_{\text{ms}}$.

Based on the above two definitions of merger point, Equations (8) and (9), the angular momentum at the merger is given by

$$J_{\text{merger}} = \begin{cases} \nu \sqrt{\frac{\mathcal{C}_2 + q\mathcal{C}_1}{(1+q)\mathcal{C}_1\mathcal{C}_2}} \frac{GM^2}{c}, & q \approx 1, \\ \nu q^{1/3} [(1+q)\mathcal{C}_2]^{-1/2} \frac{GM^2}{c}, & q < 1, \end{cases} \tag{10}$$

where we have introduced the so-called symmetric mass-ratio parameter, $\nu \equiv q/(1+q)^2$.

3.3. Mass-Energy Conservation

The conservation of mass-energy before and after the merger implies the energy released equals the mass defect of the system, i.e.,

$$E_{\text{GW}} + E_{\text{other}} = \Delta M c^2 = [M - (m_c + m_{\text{ej}} + m_d)]c^2, \tag{11}$$

where ΔM is the system's mass defect. We have also defined $E_{\text{GW}} = E_{\text{GW}}^{\text{insp}} + E_{\text{GW}}^{\text{pm}}$ the total energy emitted in GWs in the inspiral regime, $E_{\text{GW}}^{\text{insp}}$, and in the merger and post-merger phases, $E_{\text{GW}}^{\text{pm}}$. The energy E_{other} is radiated in channels different from the GW emission, e.g., electromagnetic (photons) and neutrinos.

4. A Specific Example of BNS Merger

We analyze the merger of the $1.505 + 1.404\ M_\odot$ BNS in Table 1. For these component masses, the inferred orbital separation of $a_{\text{orb}} \approx 2 \times 10^{10}$ cm and eccentricity $e = 0.45$, the merger is expected to be driven by GW radiation on a timescale [55] of

$$\tau_{\text{GW}} = \frac{c^5}{G^3} \frac{5}{256} \frac{a_{\text{orb}}^4}{\mu M^2} F(e) \approx 73.15\ \text{kyr}, \quad F(e) = \frac{48}{19} \frac{1}{g(e)^4} \int_0^e \frac{g(e)^4 (1-e^2)^{5/2}}{e(1 + \frac{121}{304}e^2)} de \approx 0.44, \tag{12}$$

where $g(e) = e^{12/19}(1 - e^2)^{-1}(1 + 121e^2/304)^{870/2299}$.

From Equations (2), (7) and (10), and the conservation Equations (1), (4) and (11), we can obtain the remnant and disk's mass as a function of the angular momentum losses, ΔJ, as well as an estimate of the energy and angular momentum released in the cataclysmic event. We use the NS structure parameters obtained for the GM1 EOS and the TM1 EOS. The total gravitational mass of the system is $M = m_1 + m_2 = 2.909\ M_\odot$, so using Equation (2), we obtain the total baryonic mass of the binary, $M_b = m_{b,1} + m_{b,2} \approx 3.184\ M_\odot$. The binary's mass fraction is $q = 0.933$, so we assume the merger starts at the contact point. With this, the angular momentum at the merger, as given by Equation (10), for the GM1 and TM1 EOS is, respectively, $J_{\text{merger}} \approx 5.65 \, GM_\odot^2/c$ and $J_{\text{merger}} \approx 5.73 \, GM_\odot^2/c$.

Figure 2 shows that the disk's mass versus the central remnant's mass for selected values of the angular momentum loss for the two EOS. The figure shows the system's final parameters lie between two limiting cases: zero angular momentum loss leading to maximal disk mass and maximal angular momentum loss leading to zero disk mass.

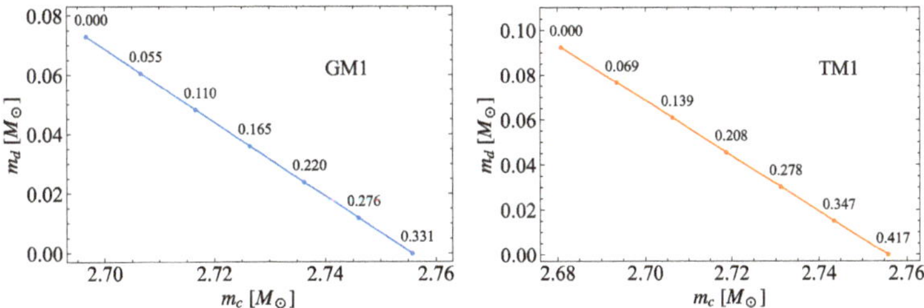

Figure 2. Disk mass versus central remnant (NS) mass. Selected values of the angular momentum loss (in units of $GM_\odot^2/c$) are shown as points. The initial BNS has a total gravitational mass of 2.909 $M_\odot$ and a mass fraction $q = 0.933$, so we assume the merger starts at the contact point. The maximum mass along the Keplerian sequence for the GM1 EOS is 2.84 $M_\odot$ and for the TM1 EOS it is 2.62 $M_\odot$ (see Table 2). Thus, for the former EOS, the central remnant is a massive fast-rotating NS, while the latter suggests a prompt collapse into a Kerr BH.

4.1. Maximal Disk Mass

We obtain the configuration corresponding to the maximum disk mass switching off angular momentum losses. Let us specialize in the GM1 EOS. By setting $\Delta J = 0$, the solution of the system of equations formed by the baryon number and angular conservation equations leads to the central remnant's mass, $m_c = 2.697 \, M_\odot$, and disk's mass, $m_d = 0.073 \, M_\odot$. This limiting case switches off the GW emission, so it also sets an upper limit to the energy released in mechanisms different than GWs. Thus, Equation (11) implies that $E_{\text{other}} = \Delta M c^2 = [M - (m_c + m_{\text{ej}} + m_d)]c^2 \approx (M - m_c - m_d)c^2 \approx 0.139 \, M_\odot c^2 \approx 2.484 \times 10^{53}$ erg of energy are carried out to infinity by a mechanism different than GWs and not accompanied by angular momentum losses.

4.2. Zero Disk Mass

The other limiting case corresponds when the angular momentum loss and the remnant mass are maximized, i.e., when no disk is formed (see Figure 2). By setting $m_d = 0$, the solution of the conservation equations leads to the maximum angular momentum loss, $\Delta J = 0.331 \, GM_\odot^2/c$, and the maximum remnant's mass, $m_c = 2.756 \, M_\odot$.

Thus, the upper limit to the angular momentum carried out by GWs is given by the maximum amount of angular momentum losses, i.e., $\Delta J_{\text{GW}} \lesssim 0.331 \, GM_\odot^2/c$. In the inspiral phase of the merger, the system releases

$$E_{\text{GW}}^{\text{insp}} \approx \frac{Gm_1 m_2}{2r_{\text{cont}}} = \frac{qC_1 C_2 M c^2}{2(1+q)(C_2 + qC_1)} + \frac{1}{2}\left[j_1|\Omega_1| + j_2|\Omega_2|\right]\frac{GM_\odot^2}{c}. \tag{13}$$

For the binary we are analyzing, $E_{\text{GW}}^{\text{insp}} \approx 0.0194 \, Mc^2 \approx 0.0563 M_\odot c^2 \approx 1.0073 \times 10^{53}$ erg. The transitional non-axisymmetric object (e.g., triaxial ellipsoid) formed immediately after the merger mainly generates these GWs, and their emission ends when the stable remnant NS is finally formed. We can model such a rotating object as a compressible ellipsoid with a polytropic EOS of index $n = 0.5$–1 [56]. The object will spin up by angular momentum loss to typical frequencies of 1.4–2.0 kHz. The energy emitted in GWs is $E_{\text{GW}}^{\text{pm}} \approx 0.0079 \, M_\odot c^2 \approx 1.404 \times 10^{52}$ erg. Therefore, the energy released in GWs is,

$E_{GW} = E_{GW}^{insp} + E_{GW}^{pm} \approx 0.0642\, M_\odot c^2 \approx 1.147 \times 10^{53}$ erg. If no disk is formed, i.e., for a U-GRF, the mass-energy defect is $\Delta M c^2 = [M - (m_c + m_{ej})]c^2 \approx (M - m_c)c^2 \approx 0.153\, M_\odot c^2 \approx 2.734 \times 10^{53}$ erg. This implies that $E_{other} = \Delta M c^2 - E_{GW} \approx 0.089\, M_\odot c^2 \approx 1.591 \times 10^{53}$ erg are released in forms of energy different than GW radiation.

Therefore, combining the above two results, we conclude that for the present merger, assuming the GM1 EOS, the merger releases $0 < E_{GW} \lesssim 1.147 \times 10^{53}$ erg in GWs and $1.591 \times 10^{53} \lesssim E_{other} < 2.484 \times 10^{53}$ erg are released in other energy forms. The energy observed in short GRBs and further theoretical analysis, including numerical simulations of the physical processes occurring during the merger, will clarify the efficiency of converting E_{other} into observable radiation. Since no BH is formed (in this GM1 EOS analysis), the assumption that the merger leads to an S-GRF suggests an efficiency lower than 10%.

We now estimate the detection efficiency of the GW radiation released by the system in the post-merger phase when angular momentum losses are maximized, i.e., in the absence of a surrounding disk. We find the root-sum-squared strain of the signal, i.e.,

$$h_{rss} = \sqrt{\int 2\left[|\tilde{h}_+|^2 + |\tilde{h}_\times|^2\right]df} \approx \frac{1}{\pi d \bar{f}}\sqrt{\frac{GE_{GW}^{pm}}{c^3}}, \tag{14}$$

where $\tilde{h}_+$ and $\tilde{h}_\times$ are the Fourier transforms of the GW polarizations, d is the distance to the source, $\bar{f}$ is the mean GW frequency in the postmerger phase. These signals are expected to be detected with a 50% of efficiency by the LIGO/Virgo pipelines [57] when $h_{rss} \sim 10^{-22}$ Hz$^{-1/2}$ [58]. For the energy release in the post-merger phase, we have $\bar{f} = 1671.77$ Hz, so these signals could be detected up to a distance of $d \approx 10$ Mpc.

5. Discussion and Conclusions

As some BdHN I and II systems remain bound after the GRB-SN event, the corresponding NS-BH and BNS systems, driven by GW radiation, will merge and lead to short GRBs. For a few minutes binary, the merger time is of the order of 10^4 year. This implies that the binaries will still be close to the long GRB site by the merger time, which implies a direct link between long and short GRBs [20].

The occurrence rate of long and short bursts, however, should differ as the SN explosion likely disrupts the binaries with long orbital periods. We are updating our previous analysis on this interesting topic reported in [59]. We refer the reader to Bianco et al. [60] for a preliminary discussion.

As a proof of concept, this article examined this unique connection between long and short GRBs predicted by the BdHN scenario, emphasizing the case of mergers of BNS left by BdHNe II. For this particular case, the simulations predict that the outcome system will be a NSs binary with the star spins anti-aligned. The application of the present theoretical framework to the analysis of other merging binaries, such as the BH-NS binaries produced by BdHN I (see [20] for a general discussion), will be addressed in a separate work.

We have carried out a numerical SPH simulation of a BdHN II occurring in a CO-NS of orbital period 4.5 min. The mass of the CO is 3.06 $M_\odot$ and that of the NS companion, 1.4 $M_\odot$. The CO is the pre-SN star obtained from a ZAMS star of $M_{zams} = 15\, M_\odot$ simulated from MESA code. The SPH simulation follows [23,43]. It computes the accretion rate onto the νNS (left by the CO core collapse) and the NS companion while the ejecta expands within the binary. For the event that left a νNS-NS eccentric binary of $1.505 + 1.404\, M_\odot$, orbital separation 2×10^{10} cm, orbital period of ≈ 15 min and eccentricity $e = 0.45$. The SN ejecta matter forms a disk around both stars with opposite spins, so we expect that the ν-NS binary will also have anti-aligned spins as well. The above parameters suggest the BNS merger leading to a short GRB occurs in ≈ 73 kyear after the BdHN II event.

Whether or not the central remnant of the BNS merger will be a Kerr BH or a massive, fast-rotating NS depends on the nuclear EOS. For instance, we have shown the GM1 EOS leads to the latter while the TM1 EOS leads to the former. As an example of the theoretical framework presented in this article, we quantify the properties of the merger using the GM1

EOS. We infer the mass of the NS central remnant and the surrounding disk as a function of the angular momentum losses. We then emphasize the merger features in the limiting cases of maximum and zero angular momentum loss, corresponding to a surrounding disk's absence or maximum mass. We estimated the maximum energy and angular momentum losses in GWs. We showed that the post-merger phase could release up to $\approx 10^{52}$ erg in ≈ 1.7 kHz GWs, and LIGO/Virgo could, in principle, detect such emissions for sources up to ≈ 10 Mpc. We assessed that up to a few 10^{53} erg of energy could be released in other forms of energy, so a $\lesssim 10\%$ of efficiency of its conversion into observable electromagnetic radiation would lead to an S-GRF.

The direct link between long and short GRB progenitors predicted by the BdHN model opens the way to exciting astrophysical developments. For instance, the relative rate of BdHNe I and II and S-GRBs and S-GRFs might give crucial information on the nuclear EOS of NSs and the CO-NS parameters. At the same time, this information provides clues for the stellar evolution path of the binary progenitors leading to the CO-NS binaries of the BdHN scenario. Although challenging because of their expected ultrashort duration, observing a U-GRF would also be relevant for constraining the EOS of NS matter. An extended analysis is encouraged, including additional BNS parameters obtained from SPH simulations of BdHNe for various CO-NS systems and nuclear EOS.

Author Contributions: Conceptualization, L.M.B., C.F., J.F.R., J.A.R. and R.R.; methodology, L.M.B., J.F.R. and J.A.R.; formal analysis, L.M.B., J.F.R. and J.A.R.; investigation, L.M.B., J.F.R. and J.A.R.; writing—original draft, L.M.B., C.F., J.F.R., J.A.R. and R.R.; writing—review and editing, L.M.B., J.F.R. and J.A.R. All authors have read and agreed to the published version of the manuscript.

Funding: L.M.B. is supported by the Vicerrectoría de Investigación y Extensión—Universidad Industrial de Santander Postdoctoral Fellowship Program No. 2023000359. J.F.R. has received funding/support from the Patrimonio Autónomo—Fondo Nacional de Financiamiento para la Ciencia, la Tecnología y la Innovación Francisco José de Caldas (MINCIENCIAS–COLOMBIA) Grant No. 110685269447 RC-80740-465-2020, Project No. 69553.

Data Availability Statement: No new data were created or analyzed in this study. Data sharing is not applicable to this article.

Abbreviations

The following abbreviations are used in this manuscript:

BdHN	Binary-driven hypernova
BH	Black hole
BNS	Binary neutron star
CO	Carbon-oxygen
EOS	Equation of state
GRB	Gamma-ray burst
GW	Gravitational wave
ISCO	Innermost stable circular orbit
NS	Neutron star
νNS	Newborn neutron star
S-GRB	Short gamma-ray burst
S-GRF	Short gamma-ray flash
SN	Supernova
U-GRB	Ultrashort gamma-ray burst
U-GRF	Ultrashort gamma-ray flash
ZAMS	Zero-age main-sequence

References

1. Mazets, E.P.; Golenetskii, S.V.; Ilinskii, V.N.; Panov, V.N.; Aptekar, R.L.; Gurian, I.A.; Proskura, M.P.; Sokolov, I.A.; Sokolova, Z.I.; Kharitonova, T.V. Catalog of cosmic gamma-ray bursts from the KONUS experiment data. I. *Astrophys. Space Sci.* **1981**, *80*, 3–83. [CrossRef]
2. Klebesadel, R.W. The durations of gamma-ray bursts. In *Gamma-Ray Bursts—Observations, Analyses and Theories*; Ho, C., Epstein, R.I., Fenimore, E.E., Eds.; Cambridge University Press: Cambridge, UK, 1992; pp. 161–168.
3. Dezalay, J.P.; Barat, C.; Talon, R.; Syunyaev, R.; Terekhov, O.; Kuznetsov, A. Short cosmic events—A subset of classical GRBs? In *American Institute of Physics Conference Series*; Paciesas, W.S., Fishman, G.J., Eds.; AIP Publishing: Melville, NY, USA, 1992; Volume 265, pp. 304–309.
4. Kouveliotou, C.; Meegan, C.A.; Fishman, G.J.; Bhat, N.P.; Briggs, M.S.; Koshut, T.M.; Paciesas, W.S.; Pendleton, G.N. Identification of two classes of gamma-ray bursts. *Astrophys. J.* **1993**, *413*, L101–L104. [CrossRef]
5. Tavani, M. Euclidean versus Non-Euclidean Gamma-Ray Bursts. *Astrophys. J.* **1998**, *497*, L21–L24. [CrossRef]
6. Goodman, J. Are gamma-ray bursts optically thick? *Astrophys. J.* **1986**, *308*, L47. [CrossRef]
7. Paczynski, B. Gamma-ray bursters at cosmological distances. *Astrophys. J.* **1986**, *308*, L43–L46. [CrossRef]
8. Eichler, D.; Livio, M.; Piran, T.; Schramm, D.N. Nucleosynthesis, neutrino bursts and gamma-rays from coalescing neutron stars. *Nature* **1989**, *340*, 126–128. [CrossRef]
9. Narayan, R.; Piran, T.; Shemi, A. Neutron star and black hole binaries in the Galaxy. *Astrophys. J.* **1991**, *379*, L17–L20. [CrossRef]
10. Woosley, S.E. Gamma-ray bursts from stellar mass accretion disks around black holes. *Astrophys. J.* **1993**, *405*, 273–277. [CrossRef]
11. Mészáros, P. Theories of Gamma-Ray Bursts. *Annu. Rev. Astron. Astrophys.* **2002**, *40*, 137. [CrossRef]
12. Piran, T. The physics of gamma-ray bursts. *Rev. Mod. Phys.* **2004**, *76*, 1143–1210. [CrossRef]
13. Galama, T.J.; Vreeswijk, P.M.; van Paradijs, J.; Kouveliotou, C.; Augusteijn, T.; Bohnhardt, H.; Brewer, J.P.; Doublier, V.; Gonzalez, J.-F.; Leibundgut, B.; et al. An unusual supernova in the error box of the γ-ray burst of 25 April 1998. *Nature* **1998**, *395*, 670–672. [CrossRef]
14. Woosley, S.E.; Bloom, J.S. The Supernova Gamma-Ray Burst Connection. *Annu. Rev. Astron. Astrophys.* **2006**, *44*, 507–556. [CrossRef]
15. Della Valle, M. Supernovae and Gamma-Ray Bursts: A Decade of Observations. *Int. J. Mod. Phys. D* **2011**, *20*, 1745–1754. [CrossRef]
16. Hjorth, J.; Bloom, J.S. The Gamma-Ray Burst—Supernova Connection. In *Gamma-Ray Bursts*; Cambridge Astrophysics Series; Kouveliotou, C., Wijers, R.A.M.J., Woosley, S., Eds.; Cambridge University Press: Cambridge, UK, 2012; Volume 51, Chapter 9; pp. 169–190.
17. Rueda, J.A.; Ruffini, R. On the Induced Gravitational Collapse of a Neutron Star to a Black Hole by a Type Ib/c Supernova. *Astrophys. J.* **2012**, *758*, L7. [CrossRef]
18. Izzo, L.; Rueda, J.A.; Ruffini, R. GRB 090618: A candidate for a neutron star gravitational collapse onto a black hole induced by a type Ib/c supernova. *Astron. Astrophys.* **2012**, *548*, L5. [CrossRef]
19. Fryer, C.L.; Rueda, J.A.; Ruffini, R. Hypercritical Accretion, Induced Gravitational Collapse, and Binary-Driven Hypernovae. *Astrophys. J.* **2014**, *793*, L36. [CrossRef]
20. Fryer, C.L.; Oliveira, F.G.; Rueda, J.A.; Ruffini, R. Neutron-Star-Black-Hole Binaries Produced by Binary-Driven Hypernovae. *Phys. Rev. Lett.* **2015**, *115*, 231102. [CrossRef]
21. Becerra, L.; Cipolletta, F.; Fryer, C.L.; Rueda, J.A.; Ruffini, R. Angular Momentum Role in the Hypercritical Accretion of Binary-driven Hypernovae. *Astrophys. J.* **2015**, *812*, 100. [CrossRef]
22. Becerra, L.; Bianco, C.L.; Fryer, C.L.; Rueda, J.A.; Ruffini, R. On the Induced Gravitational Collapse Scenario of Gamma-ray Bursts Associated with Supernovae. *Astrophys. J.* **2016**, *833*, 107. [CrossRef]
23. Becerra, L.; Ellinger, C.L.; Fryer, C.L.; Rueda, J.A.; Ruffini, R. SPH Simulations of the Induced Gravitational Collapse Scenario of Long Gamma-Ray Bursts Associated with Supernovae. *Astrophys. J.* **2019**, *871*, 14. [CrossRef]
24. Ruffini, R.; Rueda, J.A.; Muccino, M.; Aimuratov, Y.; Becerra, L.M.; Bianco, C.L.; Kovacevic, M.; Moradi, R.; Oliveira, F.G.; Pisani, G.B.; et al. On the Classification of GRBs and Their Occurrence Rates. *Astrophys. J.* **2016**, *832*, 136. [CrossRef]
25. Ruffini, R.; Muccino, M.; Kovacevic, M.; Oliveira, F.G.; Rueda, J.A.; Bianco, C.L.; Enderli, M.; Penacchioni, A.V.; Pisani, G.B.; Wang, Y.; et al. GRB 140619B: A short GRB from a binary neutron star merger leading to black hole formation. *Astrophys. J.* **2015**, *808*, 190. [CrossRef]
26. Ruffini, R.; Muccino, M.; Aimuratov, Y.; Bianco, C.L.; Cherubini, C.; Enderli, M.; Kovacevic, M.; Moradi, R.; Penacchioni, A.V.; Pisani, G.B.; et al. GRB 090510: A Genuine Short GRB from a Binary Neutron Star Coalescing into a Kerr-Newman Black Hole. *Astrophys. J.* **2016**, *831*, 178. [CrossRef]
27. Ruffini, R.; Muccino, M.; Aimuratov, Y.; Amiri, M.; Bianco, C.L.; Chen, Y.C.; Eslam Panah, B.; Mathews, G.J.; Moradi, R.; Pisani, G.B.; et al. On the universal GeV emission in short GRBs. *arXiv* **2018**, arXiv:1802.07552.
28. Aimuratov, Y.; Ruffini, R.; Muccino, M.; Bianco, C.L.; Penacchioni, A.V.; Pisani, G.B.; Primorac, D.; Rueda, J.A.; Wang, Y. GRB 081024B and GRB 140402A: Two Additional Short GRBs from Binary Neutron Star Mergers. *Astrophys. J.* **2017**, *844*, 83. [CrossRef]
29. Li, L.X.; Paczyński, B. Transient Events from Neutron Star Mergers. *Astrophys. J.* **1998**, *507*, L59–L62. [CrossRef]

30. Metzger, B.D.; Martínez-Pinedo, G.; Darbha, S.; Quataert, E.; Arcones, A.; Kasen, D.; Thomas, R.; Nugent, P.; Panov, I.V.; Zinner, N.T. Electromagnetic counterparts of compact object mergers powered by the radioactive decay of r-process nuclei. *Mon. Not. R. Astron. Soc.* **2010**, *406*, 2650–2662. [CrossRef]
31. Tanvir, N.R.; Levan, A.J.; Fruchter, A.S.; Hjorth, J.; Hounsell, R.A.; Wiersema, K.; Tunnicliffe, R.L. A 'kilonova' associated with the short-duration γ-ray burst GRB 130603B. *Nature* **2013**, *500*, 547–549. [CrossRef]
32. Berger, E.; Fong, W.; Chornock, R. An r-process Kilonova Associated with the Short-hard GRB 130603B. *Astrophys. J.* **2013**, *774*, L23. [CrossRef]
33. Metzger, B.D. Kilonovae. *Living Rev. Relativ.* **2017**, *20*, 3. [CrossRef]
34. Berger, E. Short-Duration Gamma-Ray Bursts. *Annu. Rev. Astron. Astrophys.* **2014**, *52*, 43–105. [CrossRef]
35. Tauris, T.M.; Langer, N.; Podsiadlowski, P. Ultra-stripped supernovae: Progenitors and fate. *Mon. Not. R. Astron. Soc.* **2015**, *451*, 2123–2144. [CrossRef]
36. Tauris, T.M.; Kramer, M.; Freire, P.C.C.; Wex, N.; Janka, H.T.; Langer, N.; Podsiadlowski, P.; Bozzo, E.; Chaty, S.; Kruckow, M.U.; et al. Formation of Double Neutron Star Systems. *Astrophys. J.* **2017**, *846*, 170. [CrossRef]
37. Nomoto, K.; aki Hashimoto, M. Presupernova evolution of massive stars. *Phys. Rep.* **1988**, *163*, 13–36. [CrossRef]
38. Kim, H.J.; Yoon, S.C.; Koo, B.C. Observational Properties of Type Ib/c Supernova Progenitors in Binary Systems. *Astrophys. J.* **2015**, *809*, 131. [CrossRef]
39. Yoon, S.C. Evolutionary Models for Type Ib/c Supernova Progenitors. *Publ. Astron. Soc. Aust.* **2015**, *32*, e015. [CrossRef]
40. Hamers, A.S.; Fragione, G.; Neunteufel, P.; Kocsis, B. First- and second-generation black hole and neutron star mergers in 2+2 quadruples: Population statistics. *Mon. Not. R. Astron. Soc.* **2021**, *506*, 5345–5360. [CrossRef]
41. Vynatheya, P.; Hamers, A.S. How Important Is Secular Evolution for Black Hole and Neutron Star Mergers in 2+2 and 3+1 Quadruple-star Systems? *Astrophys. J.* **2022**, *926*, 195. [CrossRef]
42. Sana, H.; de Mink, S.E.; de Koter, A.; Langer, N.; Evans, C.J.; Gieles, M.; Gosset, E.; Izzard, R.G.; Le Bouquin, J.B.; Schneider, F.R.N. Binary Interaction Dominates the Evolution of Massive Stars. *Science* **2012**, *337*, 444. [CrossRef]
43. Becerra, L.M.; Moradi, R.; Rueda, J.A.; Ruffini, R.; Wang, Y. First minutes of a binary-driven hypernova. *Phys. Rev. D* **2022**, *106*, 083002. [CrossRef]
44. Stergioulas, N.; Friedman, J.L. Comparing Models of Rapidly Rotating Relativistic Stars Constructed by Two Numerical Methods. *Astrophys. J.* **1995**, *444*, 306. [CrossRef]
45. Glendenning, N.K.; Moszkowski, S.A. Reconciliation of neutron-star masses and binding of the Lambda in hypernuclei. *Phys. Rev. Lett.* **1991**, *67*, 2414–2417. [CrossRef] [PubMed]
46. Pal, S.; Bandyopadhyay, D.; Greiner, W. Antikaon condensation in neutron stars. *Nucl. Phys. A* **2000**, *674*, 553–577. [CrossRef]
47. Sugahara, Y.; Toki, H. Relativistic mean-field theory for unstable nuclei with non-linear σ and ω terms. *Nucl. Phys. A* **1994**, *579*, 557–572. [CrossRef]
48. Cipolletta, F.; Cherubini, C.; Filippi, S.; Rueda, J.A.; Ruffini, R. Fast rotating neutron stars with realistic nuclear matter equation of state. *Phys. Rev. D* **2015**, *92*, 023007. [CrossRef]
49. Oechslin, R.; Janka, H.T.; Marek, A. Relativistic neutron star merger simulations with non-zero temperature equations of state. I. Variation of binary parameters and equation of state. *Astron. Astrophys.* **2007**, *467*, 395–409. [CrossRef]
50. Bauswein, A.; Goriely, S.; Janka, H.T. Systematics of Dynamical Mass Ejection, Nucleosynthesis, and Radioactively Powered Electromagnetic Signals from Neutron-star Mergers. *Astrophys. J.* **2013**, *773*, 78. [CrossRef]
51. Cipolletta, F.; Cherubini, C.; Filippi, S.; Rueda, J.A.; Ruffini, R. Last stable orbit around rapidly rotating neutron stars. *Phys. Rev. D* **2017**, *96*, 024046. [CrossRef]
52. Bernuzzi, S. Neutron star merger remnants. *Gen. Relativ. Gravit.* **2020**, *52*, 108. [CrossRef]
53. Chandrasekhar, S. *Ellipsoidal Figures of Equilibrium*; Dover: New York, NY, USA, 1969.
54. Shibata, M.; Taniguchi, K. Coalescence of Black Hole-Neutron Star Binaries. *Living Rev. Relat.* **2011**, *14*, 6. [CrossRef]
55. Maggiore, M. *Gravitational Waves: Volume 1: Theory and Experiments*; Oxford University Press: Oxford, UK, 2007.
56. Lai, D.; Shapiro, S.L. Gravitational radiation from rapidly rotating nascent neutron stars. *Astrophys. J.* **1995**, *442*, 259–272. [CrossRef]
57. Abbott, B.P.; Abbott, R.; Abbott, T.D.; Abernathy, M.R.; Acernese, F.; Ackley, K.; Adams, C.; Adams, T.; Addesso, P.; Adhikari, R.X.; et al. All-sky search for long-duration gravitational wave transients in the first Advanced LIGO observing run. *Class. Quantum Gravity* **2018**, *35*, 065009. [CrossRef]
58. Abbott, B.P.; Abbott, R.; Abbott, T.D.; Acernese, F.; Ackley, K.; Adams, C.; Adams, T.; Addesso, P.; Adhikari, R.X.; Adya, V.B.; et al. Search for Post-merger Gravitational Waves from the Remnant of the Binary Neutron Star Merger GW170817. *Astrophys. J.* **2017**, *851*, L16. [CrossRef]

59. Ruffini, R.; Rodriguez, J.; Muccino, M.; Rueda, J.A.; Aimuratov, Y.; Barres de Almeida, U.; Becerra, L.; Bianco, C.L.; Cherubini, C.; Filippi, S.; et al. On the Rate and on the Gravitational Wave Emission of Short and Long GRBs. *Astrophys. J.* **2018**, *859*, 30. [CrossRef]
60. Bianco, C.L.; Mirtorabi, M.T.; Moradi, R.; Rastegarnia, F.; Rueda, J.A.; Ruffini, R.; Wang, Y.; Della Valle, M.; Li, L.; Zhang, S.R. Probing electromagnetic-gravitational wave emission coincidence in type I binary-driven hypernova family of long GRBs at very-high redshift. *arXiv* **2023**, arXiv:2306.05855. [CrossRef]

universe

Review

A Short Survey of Matter-Antimatter Evolution in the Primordial Universe

Johann Rafelski *, Jeremiah Birrell, Andrew Steinmetz and Cheng Tao Yang

Department of Physics, The University of Arizona, Tucson, AZ 85721, USA
* Correspondence: johannr@arizona.edu

Abstract: We offer a survey of the matter-antimatter evolution within the primordial Universe. While the origin of the tiny matter-antimatter asymmetry has remained one of the big questions in modern cosmology, antimatter itself has played a large role for much of the Universe's early history. In our study of the evolution of the Universe we adopt the position of the standard model Lambda-CDM Universe implementing the known baryonic asymmetry. We present the composition of the Universe across its temperature history while emphasizing the epochs where antimatter content is essential to our understanding. Special topics we address include the heavy quarks in quark-gluon plasma (QGP), the creation of matter from QGP, the free-streaming of the neutrinos, the vanishing of the muons, the magnetism in the electron-positron cosmos, and a better understanding of the environment of the Big Bang Nucleosynthesis (BBN) producing the light elements. We suggest but do not explore further that the methods used in exploring the early Universe may also provide new insights in the study of exotic stellar cores, magnetars, as well as gamma-ray burst (GRB) events. We describe future investigations required in pushing known physics to its extremes in the unique laboratory of the matter-antimatter early Universe.

Keywords: particles; plasmas and electromagnetic fields in cosmology; quarks to cosmos

1. Timeline of Particles and Plasmas in the Universe

1.1. Guide to 130 GeV > T > 20 keV

This survey of the early Universe begins with quark-gluon plasma (QGP) at a temperature of $T = 130$ GeV. It then ends at a temperature of $T = 20$ keV with the electron-positron epoch which was the final phase of the Universe to contain significant quantities of antimatter. This defines the "short" $t \approx 1/2$ h time-span that will be covered. This work presumes that the Universe is homogeneous and that in our casual domain, the Universe's baryon content is matter dominated. Our work is rooted in the Universe as presented by Lizhi Fang and Remo Ruffini [1–3]. Within the realm of the Standard Model, we coherently connect the differing matter-antimatter plasmas as each transforms from one phase into another.

A more detailed description of particles and plasmas follows in Section 1.2. We have adopted the standard ΛCDM model of a cosmological constant (Λ) and cold dark matter (CDM) where the Universe undergoes dynamical expansion as described in the Friedmann-Lemaître-Robertson-Walker (FLRW) metric. The contemporary history of the Universe in terms of energy density as a function of time and temperature is shown in Figure 1. The Universe's past is obtained from integrating backwards the proposed modern composition of the Universe which contains 69% dark energy, 26% dark matter, 5% baryons, and <1% photons and neutrinos in terms of energy density. The method used to obtain these results are found in Section 1.3.

Citation: Rafelski, J.; Birrell, J.; Steinmetz, A.; Yang, C.T. A Short Survey of Matter-Antimatter Evolution in the Primordial Universe. *Universe* **2023**, *9*, 309. https://doi.org/10.3390/universe9070309

Academic Editor: Stefano Profumo

Received: 28 April 2023
Revised: 16 June 2023
Accepted: 19 June 2023
Published: 27 June 2023

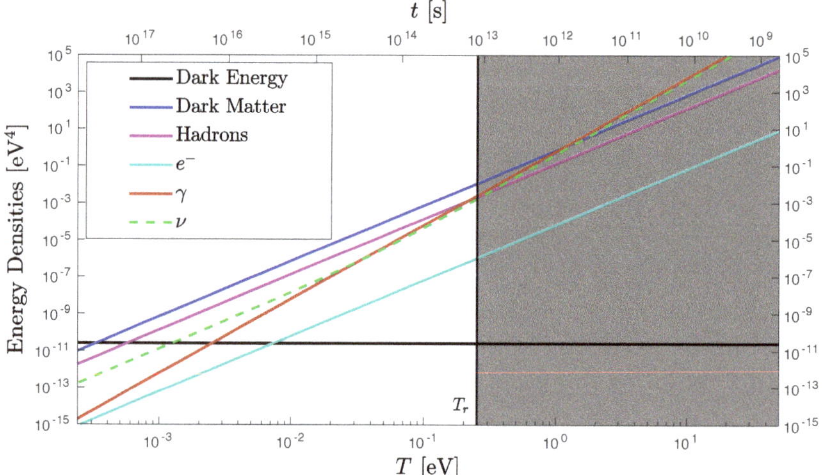

Figure 1. Contemporary and recent Universe composition: In this example we assumed present day composition to be 69% dark energy, 26% dark matter, 5% baryons, <1% photons and neutrinos. The dashed line shows how introduction of 2×0.1 eV mass in two of the three neutrinos impacts the energy density evolution (Neutrino mass choice is just for illustration. Other values are possible). The recombination temperature $T_r \approx 0.25$ eV delimits the era when the Universe was opaque shown as the shaded region.

After the general overview, we take the opportunity to enlarge in some detail our more recent work in special topics. In Section 2, we describe the chemical potentials of the QGP plasma species leading up to hadronization, Hubble expansion of the QGP plasma, and the abundances of heavy quarks. In Section 3 we discuss the formation of matter during hadronization, the role of strangeness, and the unique circumstances which led to pions remaining abundant well after all other hadrons were diluted or decayed. We review the roles of muons and neutrinos in the leptonic epoch in Section 4. The $e^{\pm}$ plasma epoch is described in Section 5 which is the final stage of the Universe where antimatter played an important role. Here we introduce the statistical physics description of electrons and positron gasses, their relation to the baryon density, and the magnetization of the $e^{\pm}$ plasma prior to the disappearance of the positrons shortly after Big Bang Nucleosynthesis (BBN). A more careful look at the effect of the dense $e^{\pm}$ plasma on BBN is underway. One interesting feature of having an abundant $e^{\pm}$ plasma is the possibility of magnetization in the early Universe which we consider in Section 5.2. We introduce in this work the spin magnetic moment polarization for the first time in the context of cosmology. We address this using spin-magnetization and mean-field theory where all the spins respond to the collective bulk magnetism self generated by the plasma. We stop our survey at a temperature of $T = 20$ keV with the disappearance of the positrons signifying the end of antimatter dynamics at cosmological scales.

This primordial Universe is a plasma physics laboratory with unique properties not found in terrestrial laboratories or stellar environments due to the high amount of antimatter present. We suggest in Section 6 areas requiring further exploration including astrophysical systems where positron content is considerable and the possibility for novel compact objects with persistent positron content is discussed. While the disappearance of baryonic matter is well described in the literature, it has not always been appreciated how long the leptonic ($\bar{\mu} = \mu^{+}$ and $\bar{e} = e^{+}$) antimatter remains a significant presence in the Universe's evolutionary history. We show that the $e^{\pm}$ epoch is a prime candidate to resolve several related cosmic mysteries such as early Universe matter in-homogeneity and the origin of cosmic magnetic fields. While the plasma epochs of the early Universe are in our long gone past, plasmas which share features with the primordial Universe might possibly exist in the contemporary Universe today. Such extraordinary stellar objects could poses properties dynamics relevant to gamma-ray burst (GRB) [4–7], black holes [8–10] and neutron stars (magnetars) [11,12].

1.2. The Five Plasma Epochs

At an early time in the standard cosmological model, the Universe began as a fireball, filling all space, with extremely high temperature and energy density [13]. Our domain of the present day Universe originated from an ultra-relativistic plasma which contained almost a perfect symmetry between matter and antimatter except for a small discrepancy of one part in 10^9 which remains a mystery today. There are two general solutions of this problem both of which suppose that the Universe's initial conditions were baryon-antibaryon number symmetric in order to avoid 'fine-tuning' to a specific value:

A Case of baryonic number (charge) conservation: In order to separate space domains in which either matter or antimatter is albeit very slightly dominant we need a 'force' capable of dynamically creating this matter-antimatter separation. This requires that two of the three Sakharov [14,15] conditions be fulfilled:

1. Violation of CP-invariance allowing to distinguish matter from antimatter
2. Non-stationary conditions in absence of local thermodynamic equilibrium

Other than very distant antimatter domains [16] the missing antimatter could be perhaps 'stored' in a compact structure [17–19].

B There is no known cause for baryon charge conservation. Therefore it is possible to consider the full Sakharov model with

3. Absence of baryonic charge conservation

Allowing the dynamical formation of the uniform matter-antimatter asymmetry typically occurring prior to the epoch governed by physics confirmed by current experiment to which environs we restrict this short survey. A well studied example is the Affleck-Dine mechanism [20].

Very early formation of baryon asymmetry is further supported by the finding that the known CP-violation in the Standard Model's weak sector is insufficient to explain in quantitative terms the baryon asymmetry [21]. However, baryon asymmetry could develop at a later stage in Universe evolution. We show in this review that this remains a topic deserving further investigation. In this work we take a homogeneous prescribed baryon asymmetry obtained from observed baryon to photon ratio in the Universe. Additional comments on the situation in the context of non-equilibria processes are made in Section 2.2, at the end of Section 4.2, and in Section 6.

The primordial hot Universe fireball underwent several practically adiabatic phase changes which dramatically evolved its bulk properties as it expanded and cooled. We present an overview Figure 2 of particle families across all epochs in the Universe, as a function of temperature and thus time. The comic plasma, after the electroweak symmetry breaking epoch and presumably inflation, occurred in the early Universe in the following sequence:

1. **Primordial quark-gluon plasma**: At early times when the temperature was between $130\,\text{GeV} > T > 150\,\text{MeV}$ we have the building blocks of the Universe as we know them today, including the leptons, vector bosons, and all three families of deconfined quarks and gluons which propagated freely. As all hadrons are dissolved into their constituents during this time, strongly interacting particles u, d, s, t, b, c, g controlled the fate of the Universe. Here we will only look at the late-stage evolution at around 150 MeV.

2. **Hadronic epoch**: Around the hadronization temperature $T_h \approx 150\,\text{MeV}$, a phase transformation occurred forcing the strongly interacting particles such as quarks and gluons to condense into confined states [22]. It is here where matter as we know it today forms and the Universe becomes hadronic-matter dominated. In

the temperature range $150\,\text{MeV} > T > 20\,\text{MeV}$ the Universe is rich in physics phenomena involving strange mesons and (anti)baryons including (anti)hyperon abundances [23,24].

3. **Lepton-photon epoch**: For temperature $10\,\text{MeV} > T > 2\,\text{MeV}$, the Universe contained relativistic electrons, positrons, photons, and three species of (anti)neutrinos. Muons vanish partway through this temperature scale. In this range, neutrinos were still coupled to the charged leptons via the weak interaction [25,26]. During this time the expansion of the Universe is controlled by leptons and photons almost on equal footing.

4. **Final antimatter epoch**: After neutrinos decoupled and become free-streaming, referred to as neutrino freeze-out, from the cosmic plasma at $T = 2\,\text{MeV}$, the cosmic plasma was dominated by electrons, positrons, and photons. We have shown in [27] that this plasma existed until $T \approx 0.02\,\text{MeV}$ such that BBN occurred within a rich electron-positron plasma. This is the last time the Universe will contain a significant fraction of its content in antimatter.

5. **Moving towards a matter dominated Universe**: The final major plasma stage in the Universe began after the annihilation of the majority of $e^{\pm}$ pairs leaving behind a residual amount of electrons determined by the baryon asymmetry in the Universe and charge conservation. The Universe was still opaque to photons at this point and remained so until the recombination period at $T \approx 0.25\,\text{eV}$ starting the era of observational cosmology with the CMB. This final epoch of the primordial Universe will not be described in detail here, but is well covered in [28].

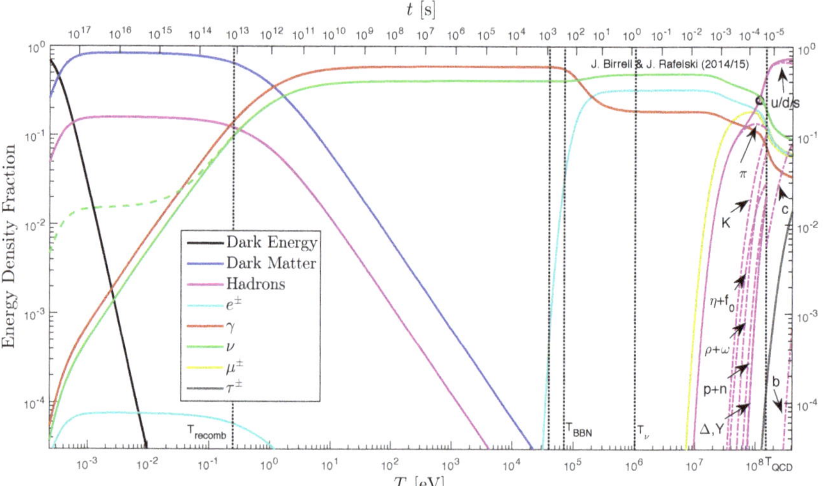

Figure 2. Normalized Universe constituent matter and radiation components Ω_i are evolved over cosmological timescales (top scale, bottom scale is temperature T) from contemporary observational cosmology to the QGP epoch of the Universe. Vertical lines denote transitions between distinct epochs. Solid neutrino (green) line shows contribution of massless neutrinos, while the dashed line shows 1 massless and $2 \times 0.1\,\text{eV}$ neutrinos (Neutrino mass choice is just for illustration. Other values are possible).

Each plasma outlined above contributes to the thermal behavior of the Universe over time. This is illustrated in Figure 3 where the fractional drop in temperature during each plasma transformation is plotted. Each subsequent plasma lowers the available degrees of freedom (as the particle inventory is whittled away) as the Universe cools [29,30]. Each drop in degrees of freedom represents entropy being pumped into the photons as entropy is conserved (up until local gravitational processes become relevant) in an expanding

Universe. As there are no longer degrees of freedom to consume, thereby reheating the photon field further, the fractional temperature remains constant today.

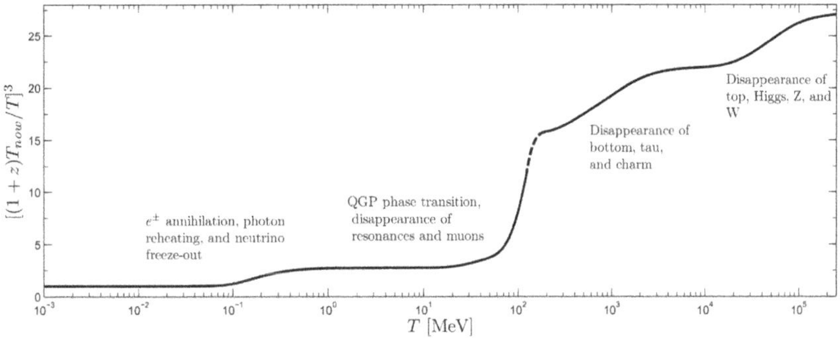

Figure 3. The evolution of the photon reheating (black line) process in terms of fractional temperature change in the Universe. Figure adapted from [29]. The dashed portion is a qualitative description subject to the exact model of QGP hadronization.

In Figure 2 we begin on the right at the end of the QGP era. The first dotted vertical line shows the QGP phase transition and hadronization, near $T = 150$ MeV. The hadron era proceeds with the disappearance of muons, pions, and heavier hadrons. This constitutes a reheating period, with energy and entropy from these particles being transferred to the remaining $e^{\pm}$, photon, neutrino plasma. The black circle near $T = 115$ MeV denotes our change from $2 + 1$-flavor lattice QCD [31–33] data for the hadron energy density, taken from Borsanyi et al. [34,35], to an ideal gas model [36] at lower temperature. We note that the hadron ideal gas energy density matches the lattice results to less than a percent at $T = 115$ MeV [37].

To the right of the QGP transition region, the solid hadron line shows the total energy density of quarks and gluons. From top to bottom, the dot-dashed hadron lines to the right of the transition show the energy density fractions of $2 + 1$-flavor (u,d,s) lattice QCD matter (almost indistinguishable from the total energy density), charm, and bottom (both in the ideal gas approximation). To the left of the transition the dot-dashed lines show the pion, kaon, $\eta + f_0$, $\rho + \omega$, nucleon, Δ, and Υ contributions to the energy fraction.

Continuing to the second vertical line at $T = \mathcal{O}\,(1\text{ MeV})$, we come to the annihilation of $e^{\pm}$ and the photon reheating period. Notice that only the photon energy density fraction increases, as we assume that neutrinos are already decoupled at this time and hence do not share in the reheating process, leading to a difference in photon and neutrino temperatures. This is not strictly correct but it is a reasonable simplifying assumption for the current purpose; see [25,38–40]. We next pass through a long period, from $T = \mathcal{O}\,(1\text{ MeV})$ until $T = \mathcal{O}\,(1\text{ eV})$, where the energy density is dominated by photons and free-streaming neutrinos. BBN occurs in the approximate range $T = 40$–70 keV and is indicated by the next two vertical lines in Figure 2. It is interesting to note that, while the hadron fraction is insignificant at this time, there is still a substantial background of $e^{\pm}$ pairs during BBN (see Section 5.1).

We then come to the beginning of the matter dominated regime, where the energy density is dominated by the combination of dark matter and baryonic matter. This transition is the result of the redshifting of the photon and neutrino energy, $\rho \propto a^{-4} \propto T^4$, whereas for non-relativistic matter $\rho \propto a^{-3} \propto T^3$. Recombination and photon decoupling occurs near the transition to the matter dominated regime, denoted by the (Figure 2) vertical line at $T = 0.25$ eV.

Finally, as we move towards the present day CMB temperature of $T_{\gamma,0} = 0.235$ meV on the left hand side, we have entered the dark energy dominated regime. For the present day values, we have used the energy densities proscribed by the Planck parameters [41] using Equation (14) and zero Universe spatial curvature. The photon energy density is fixed by the CMB temperature $T_{\gamma,0}$ and the neutrino energy density is fixed by $T_{\gamma,0}$ along with the photon to neutrino temperature ratio and neutrino masses. Both constitute $<1\%$ of the current energy budget.

The Universe evolution and total energy densities were computed using massless neutrinos, but for comparison we show the energy density of massive neutrinos in the dashed green line. For the dashed line we used two neutrino flavors with masses $m_\nu = 0.1$ eV and one massless flavor. Note that the inclusion of neutrino mass causes the leveling out of the neutrino energy density fraction during the matter dominated period, as compared to the continued redshifting of the photon energy.

1.3. The Lambda-CDM Universe

Here we provide background on the standard ΛCDM cosmological (FLRW-Universe) model that is used in the computation of the composition of the Universe over time. We use the spacetime metric with metric signature $(+1, -1, -1, -1)$ in spherical coordinates

$$ds^2 = c^2 dt^2 - a^2(t)\left[\frac{dr^2}{1 - kr^2} + r^2(d\theta^2 + \sin^2(\theta)d\phi^2)\right] \tag{1}$$

characterized by the scale parameter $a(t)$ of a spatially homogeneous Universe. The geometric parameter k identifies the Gaussian geometry of the spacial hyper-surfaces defined by co-moving observers. Space is a Euclidean flat-sheet for the observationally preferred value $k = 0$ [28,41,42]. In this case it can be more convenient to write the metric in rectangular coordinates

$$ds^2 = c^2 dt^2 - a^2(t)\left[dx^2 + dy^2 + dz^2\right]. \tag{2}$$

We will work in units where $\hbar = 1$, $c = 1$.

The global Universe dynamics can be characterized by two quantities: the Hubble parameter H, a strongly time dependent quantity on cosmological time scales, and the deceleration parameter q:

$$H(t)^2 \equiv \left(\frac{\dot{a}}{a}\right)^2 = \frac{8\pi G_N}{3}\rho_{tot}, \tag{3}$$

$$\frac{\ddot{a}}{a} = -qH^2, \qquad q \equiv -\frac{a\ddot{a}}{\dot{a}^2}, \qquad \dot{H} = -H^2(1 + q), \tag{4}$$

where G_N is the Newtonian gravitational constant and ρ_{tot} is the energy density of the Universe and composed of the various energy densities in the Universe. The deceleration parameter q is defined in terms of the second derivative of the scale parameter.

In Figure 4 Left we illustrate the late stage evolution of the parameters H and q given in Equations (3) and (4) compared to temperature. This illustrates how the Universe evolves according to the Friedmann Equations (3) and (4) above. The deceleration begins radiation dominated with $q = 1$ and then transitions to matter dominated $q = 1/2$. Within the ΛCDM model the contemporary Universe is undergoing a transition from matter dominated to dark energy dominated, where the deceleration would settle on the asymptotic value of $q = -1$ [29]. However, several alternate models: phantom energy [43], Chaplygin gas [44], or more generally dynamic (spatially and/or time dependent) dark energy [45] cannot be excluded in absence of strong evidence for the constancy of dark energy.

Within the ΛCDM model only usual forms of energy are relevant before recombination epoch, see Figure 2. Any alternate model can be thus constrained by understanding precisely the evolution of the Universe prior to this epoch. Part of the program of this survey is to connect the late stage evolution to the very early Universe during and prior to BBN accounting for the unexpectedly considerable antimatter content.

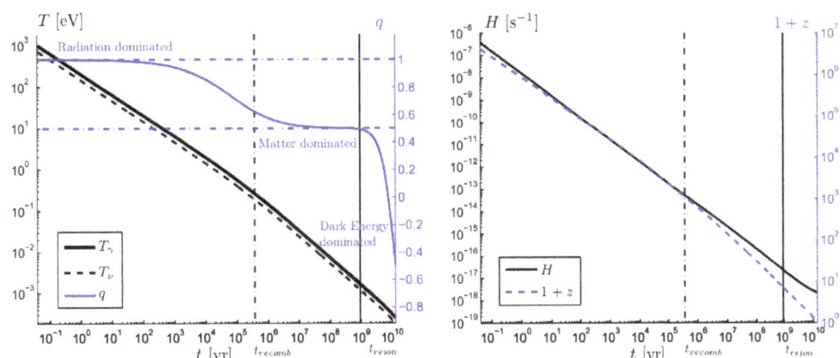

Figure 4. Left: The numerically solved later $t > 10^{-1}$ yr evolution of photon and neutrino background temperatures T_γ, T_ν (black and black dashed lines) and the deceleration parameter q (thin blue line) over the lifespan of the Universe. **Right**: The evolution of the Hubble parameter $1/H$ (black line) and redshift z (blue dashed line) which is related to the scale parameter $a(t)$. Figure adapted from [29].

The current tension in Hubble parameter measurements [46–48] might benefit from closer inspection of these earlier denser periods should these contribute to modification of the conventional model of Universe expansion. We further note that the JWST has recently discovered that galaxy formation began earlier than predicted which requires reevaluation of early Universe matter inhomogeneities [49]. Figure 4 Right shows the close relationship between the redshift z and the Hubble parameter. Deviations separating the two occur from the transitions which changed the deceleration value.

The Einstein equations with a cosmological constant Λ corresponding to dark energy are:

$$G^{\mu\nu} = R^{\mu\nu} - \left(\frac{R}{2} + \Lambda\right)g^{\mu\nu} = 8\pi G_N T^{\mu\nu}, \quad R = g_{\mu\nu}R^{\mu\nu}. \tag{5}$$

The homogeneous and isotropic symmetry considerations imply that the stress energy tensor is determined by an energy density and an isotropic pressure

$$T^{\mu}_{\nu} = \text{diag}(\rho, -P, -P, -P). \tag{6}$$

It is common to absorb the Einstein cosmological constant Λ into the energy and pressure

$$\rho_\Lambda = \frac{\Lambda}{8\pi G_N}, \quad P_\Lambda = -\frac{\Lambda}{8\pi G_N} \tag{7}$$

and we implicitly consider this done from now on.

Two dynamically independent Friedmann equations [50] arise using the metric Equation (1) in Equation (5):

$$\frac{8\pi G_N}{3}\rho = \frac{\dot{a}^2 + k}{a^2} = H^2\left(1 + \frac{k}{\dot{a}^2}\right), \quad \frac{4\pi G_N}{3}(\rho + 3P) = -\frac{\ddot{a}}{a} = qH^2. \tag{8}$$

We can eliminate the strength of the interaction, G_N, solving both these equations for $8\pi G_N/3$, and equating the result to find a relatively simple constraint for the deceleration parameter:

$$q = \frac{1}{2}\left(1 + 3\frac{P}{\rho}\right)\left(1 + \frac{k}{\dot{a}^2}\right). \tag{9}$$

For a spatially flat Universe, $k = 0$, note that in a matter-dominated era where $P/\rho << 1$ we have $q \simeq 1/2$; for a radiative Universe where $3P = \rho$ we find $q = 1$; and in a dark energy Universe in which $P = -\rho$ we find $q = -1$. Spatial flatness is equivalent to the assertion that the energy density of the Universe equals the critical density

$$\rho = \rho_{\text{crit}} \equiv \frac{3H^2}{8\pi G_N}.$$ (10)

The CMB power spectrum is sensitive to the deceleration parameter and the presence of spatial curvature modifies q. The Planck results [28,41,42] constrain the effective curvature energy density fraction,

$$\Omega_K \equiv 1 - \rho/\rho_{\text{crit}},$$ (11)

to

$$|\Omega_K| < 0.005.$$ (12)

This indicates a nearly flat Universe which is spatially Euclidean. We will work within an exactly spatially flat cosmological model, $k = 0$. As must be the case for any solution of Einstein's equations, Equation (8) implies that the energy momentum tensor of matter is divergence free:

$$T^{\mu\nu}{}_{;\nu} = 0 \Rightarrow -\frac{\dot{\rho}}{\rho + P} = 3\frac{\dot{a}}{a} = 3H.$$ (13)

A dynamical evolution equation for $\rho(t)$ arises once we combine Equation (13) with Equation (8), eliminating H. Given an equation of state $P(\rho)$, solutions of this equation describes the dynamical evolution of matter in the Universe. In practice, we evolve the system in both directions in time. On one side, we start in the present era with the energy density fractions fit by the central values found in Planck data [41]

$$H_0 = 67.4\,\text{km/s/Mpc}, \quad \Omega_b = 0.05, \quad \Omega_c = 0.26, \quad \Omega_\Lambda = 0.69,$$ (14)

and integrate backward in time. On the other hand, we start in the QGP era with an equation of state determined by an ideal gas of SM particles, combined with a perturbative QCD equation of state for quarks and gluons [35], and integrate forward in time. As the Universe continues to dilute from dark energy in the future, the cosmic equation of state will become well approximated by the de Sitter inflationary metric which is a special case of FLRW.

2. QGP Epoch

2.1. Conservation Laws in QGP

During the first $\Delta t \approx 30\,\mu$s after the Big Bang, the early Universe is a hot soup that containing the elementary primordial building blocks of matter and antimatter [13]. In particular it contained the light quarks which are now hidden in protons and neutrons. Beyond this there were also electrons, photons, neutrinos, and massive strange and charm quarks. These interacting particle species were kept in chemical and thermal equilibrium with one another. Gluons which mediated the color interaction are very abundant as well. This primordial phase lasted as long as the temperature of the Universe was more than 110,000 times than the expected temperature $T_\odot = 1.36$ keV (1.58×10^7 K) at the center of the Sun [51].

The conditions in the early Universe and those created in relativistic collisions of heavy atomic nuclei differ somewhat: whereas the primordial quark-gluon plasma survives for about 25 μs in the Big Bang, the comparable extreme conditions created in ultra-relativistic nuclear collisions are extremely short-lived [52] on order of 10^{-23} s. As a consequence of the short lifespan of laboratory QGP in heavy-ion collisions [53,54], they are not subject to the same weak interaction dynamics [55] as the characteristic times for weak processes are too lengthy [56]. Therefore our ability to recreate the conditions of the primordial QGP are limited due to the relativistic explosive disintegration of the extremely hot dense relativistic 'fireballs' created in modern accelerators. This disparity is seen in Figure 5 where the chemical potential of QGP $\mu_q = \mu_B/3$ [57] for various values of entropy-per-baryon s/b relevant to relativistic particle accelerators are plotted alongside the evolution of the cosmic hadronic plasma chemical potential. The confinement transition boundary (red line in Figure 5) was calculated using a parameters obtained from [58] in agreement with lattice results [59]. The QGP precipitates hadrons in the cosmic fluid at a far higher entropy ratio than those accessible by terrestrial means and the two manifestations of QGP live far away from each other on the QCD phase diagram [60].

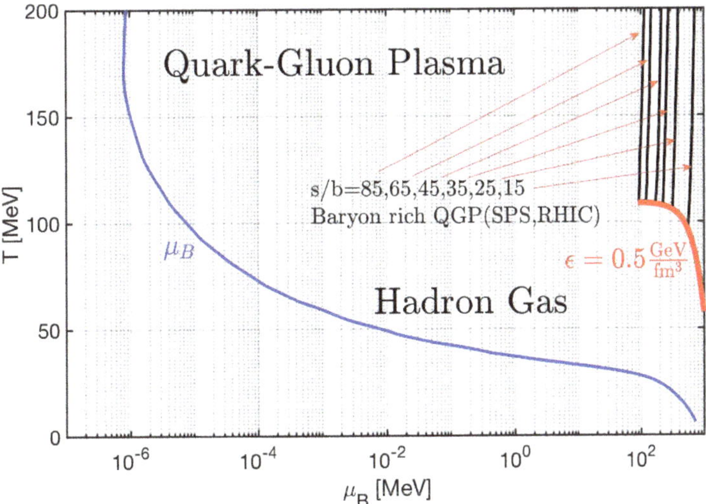

Figure 5. The evolution of the cosmic baryon chemical potential μ_B after hadronization (blue line). Curves for QGP (thin black line) created in terrestrial accelerators for differing entropy-per-baryon s/b values are included [57]. The boundary (red line) where QGP condenses into hadrons is illustrated at an energy density of $0.5\,\text{GeV}/\text{fm}^3$ as determined through lattice computation [59].

The work of Fromerth et al. [23] allows us to parameterize the chemical potentials μ_d, μ_e, and μ_ν during this epoch as they are the lightest particles in each main thermal category: quarks, charged leptons, and neutral leptons. The quark chemical potential is determined by the following three constraints [23]:

1. Electric charge neutrality $Q = 0$, given by

$$\frac{Q}{V} = n_Q \equiv \sum_f Q_f\, n_f(\mu_f, T) = 0 \tag{15}$$

where Q_f is the charge and n_f is the numerical density of each species f. Q is a conserved quantity in the Standard Model under global $U(1)_{EM}$ symmetry. This is summed is over all particles present in the QGP epoch.

2. Baryon number and lepton number neutrality $B - L = 0$, given by

$$\frac{B - L}{V} = n_B - n_L \equiv \sum_f (B_f - L_f) n_f(\mu_f, T) = 0 \tag{16}$$

where L_f and B_f are the lepton and baryon number for the given species f. This condition is phenomenologically motivated by baryogenesis and is exactly conserved in the Standard Model under global $U(1)_{B-L}$ symmetry. We note many Beyond-Standard-Model (BSM) models also retain this as an exact symmetry though Majorana neutrinos do not.

3. The entropy-per-baryon density ratio s/n_B is a constant and can be written as

$$\frac{S}{B} = \frac{s}{n_B} = \frac{\sum_f s_f(\mu_f, T)}{\sum_f B_f n_f(\mu_f, T)} = \text{const} \tag{17}$$

where s_f is the entropy density of given species f. As the expanding Universe remains in thermal equilibrium, the entropy is conserved within a co-moving volume. The baryon number within a co-moving volume is also conserved. As both quantities dilute with $1/a(t)^3$ within a normal volume, the ratio of the two is constant. This constraint does not become broken until spatial inhomogeneities from gravitational attraction becomes significant, leading to increases in local entropy.

At each temperature T, the above three conditions form a system of three coupled, nonlinear equations of the three chosen unknowns (here we have μ_d, μ_e, and μ_ν). In Figure 6 we present numerical solutions to the conditions Equations (15)–(17) and plot the chemical potentials as a function of time. As seen in the figure, the three potentials are in alignment during the QGP phase until the hadronization epoch where the down quark chemical potential diverges from the leptonic chemical potentials before reaching an asymptotic value at late times. This asymptotic value is given as approximately $\mu_q \approx m_N/3$ the mass of the nucleons and represents the confinement of the quarks into the protons and neutrons at the end of hadronization.

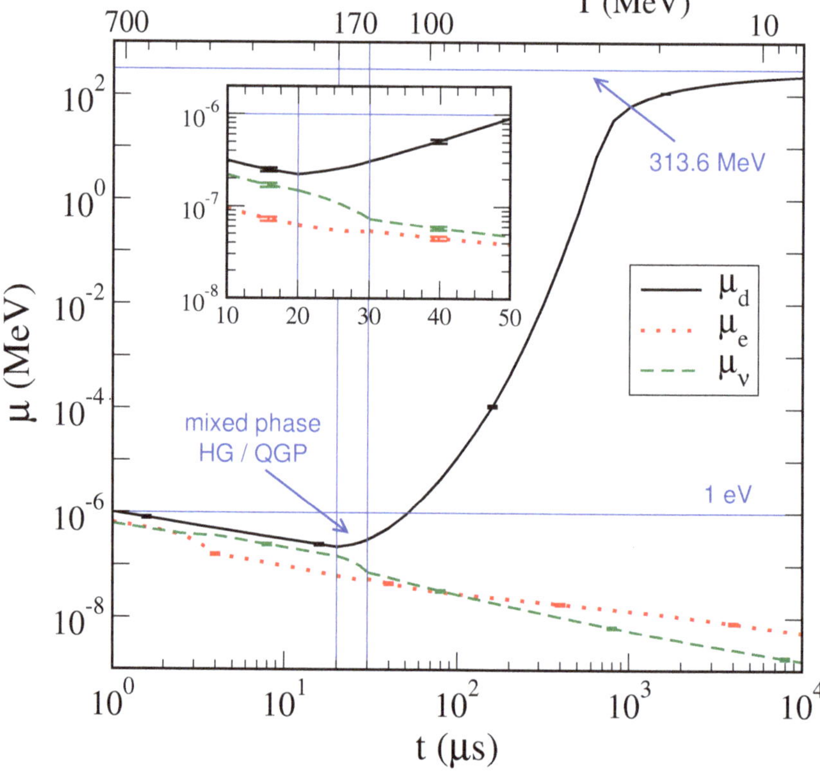

Figure 6. Plot of the down quark chemical potential (black), electron chemical potential (dotted red) and neutrino chemical potential (dashed green) as a function of time. These are 2003 unpublished results of Fromerth & Rafelski [61]; also presented in Ref. [62]).

This asymptotic limit is also shown in Figure 7 where we present the down quark chemical potential for different values of the entropy-to-baryon ratio. While the s/n_B ratio has large consequences for the plasma at high temperatures, the chemical potential is insensitive to this parameter at low temperatures the degrees of freedom are dominated by the remaining baryon number rather than the thermal degrees of freedom of the individual quarks. Therefore the entropy to baryon value today greatly controls the quark content when the Universe was very hot. We note that the distribution of quarks in the QGP plasma does not remain fixed to the Fermi-Dirac distribution for thermal and entropic equilibrium. The quark partition function is instead

$$\ln \mathcal{Z}_{\text{quarks}} = \sum_q \ln\left(1 + Y_q(t)e^{-\beta E_q}\right), \qquad Y_q(t) = \gamma_q(t)\lambda_q \qquad q = u, d, c, s, t, b, \tag{18}$$

which is summed over all quarks and their quantum numbers. In Equation (18), λ_q is the quark fugacity while $\gamma_q(t)$ is the temporal inhomogeneity of the population distribution [62]. The product of the two $Y_q(t) = \gamma_q(t)\lambda_q$ is then defined as the generalized fugacity for the species. Because of nuclear reactions, these distributions populate and depopulate over time which pulls the gas off entropic equilibrium while retaining temperature T with the rest of the Universe [58]. When $\gamma \neq 1$, the entropy of the quarks is no longer minimized. As entropy in the cosmic expansion is conserved overall, this means the entropy gain or loss is then related to the entropy moving between the quarks or its products.

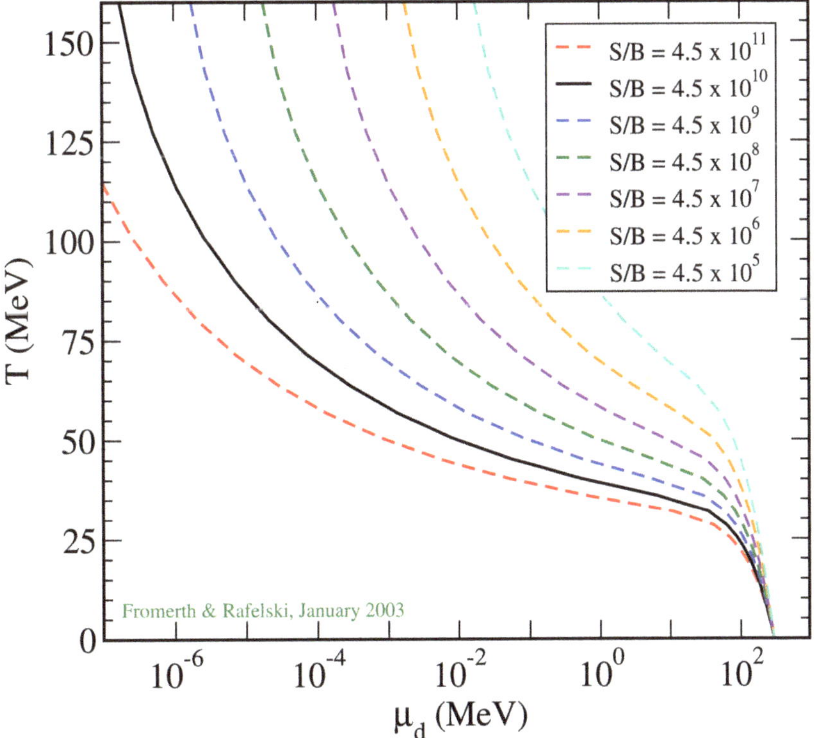

Figure 7. Plot of the down quark chemical potential μ_d as a function of temperature for differing values of entropy-per-baryon S/B ratios (2003 unpublished, Fromerth & Rafelski [62]).

In practice, the generalized fugacity is $Y = 1$ during the QGP epoch as the quarks in early Universe remained in both thermal and entropic equilibrium. This is because the Universe's expansion was many orders of magnitude slower than the process reaction and decay timescales [58]. However near the hadronization temperature, heavy quarks abundance and deviations from chemical equilibrium have not yet been studied in great detail. We show in Section 2.2 and [63] that the bottom quarks can deviate from chemical equilibrium $\gamma \neq 1$ by breaking the detailed balance between reactions of the quarks.

2.2. Heavy Flavor: Bottom and Charm in QGP

In the QGP epoch, up and down (u, d) (anti)quarks are effectively massless and remain in equilibrium via quark-gluon fusion. Strange (s) (anti)quarks are in equilibrium via weak, electromagnetic, and strong interactions until $T \sim 12$ MeV [24]. In this section, we focus on the heavier charm and bottom (c, b) (anti)quarks. In primordial QGP, the bottom and charm quarks can be produced from strong interactions via quark-gluon pair fusion processes and disappear via weak interaction decays. For production, we have the following processes

$$q + q \longrightarrow b + \bar{b}, \qquad q + q \longrightarrow c + \bar{c}, \tag{19}$$
$$g + g \longrightarrow b + \bar{b}, \qquad g + g \longrightarrow c + \bar{c}, \tag{20}$$

for bottom and charm and

$$b \longrightarrow c + l + \nu_l, \qquad b \longrightarrow c + q + \bar{q} \tag{21}$$
$$c \longrightarrow s + l + \nu_l, \qquad c \longrightarrow s + q + \bar{q} \tag{22}$$

for their decay. A detailed calculation of production and decay rate can be found in [63].

In the early Universe within the temperature range $130\,\text{GeV} > T > 150\,\text{MeV}$ we have the following particles: photons, 8_c-gluons, $W^{\pm}$, Z^0, three generations of 3_c-quarks and leptons in the primordial QGP. The Hubble parameter can be written as the sum of particle energy densities ρ_i for each species

$$H^2 = \frac{8\pi G_N}{3} \left(\rho_\gamma + \rho_{\text{lepton}} + \rho_{\text{quark}} + \rho_{g, W^{\pm}, Z^0} \right), \tag{23}$$

where G_N is Newton's constant of gravitation. Ultra-relativistic particles (which are effectively massless) and radiation dominate the speed of expansion.

The Universe's characteristic expansion time constant $1/H$ is seen in Figure 8 (both Top and Bottom figures). The (top) figure plots the relaxation time for the production and decay of charm quarks as a function of temperature. For the entire duration of QGP, the Hubble time is larger than the decay lifespan and production times of the charm quark. Therefore, the heavy charm quark remains in equilibrium as its processes occur faster than the expansion of the Universe. Additionally, the charm quark production time is faster than the charm quark decay. The faster quark-gluon pair fusion keeps the charm in chemical equilibrium up until hadronization. After hadronization, charm quarks form heavy mesons that decay into multi-particles quickly. Charm content then disappears from the Universe's particle inventory.

In Figure 8 Bottom we plot the relaxation time for production and decay of the bottom quark with different masses as a function of temperature. It shows that both production and decay are faster than the Hubble time $1/H$ for the duration of QGP. Unlike charm quarks however, the relaxation time for bottom quark production intersects with bottom quark decay at a temperatures dependant on the mass of the bottom. This means that the bottom quark decouples from the primordial plasma before hadronization as the production process slows down at low temperatures. The speed of weak interaction decays then dilutes bottom quark content of the QGP plasma pulling the distribution off equilibrium with $Y \neq 1$ (see Equation (18)) in the temperature domain below the crossing point, but before hadronization. All of this occurs with rates faster than Hubble expansion and thus as the Universe expands, the system departs from a detailed chemical balance rather than thermal freezeout.

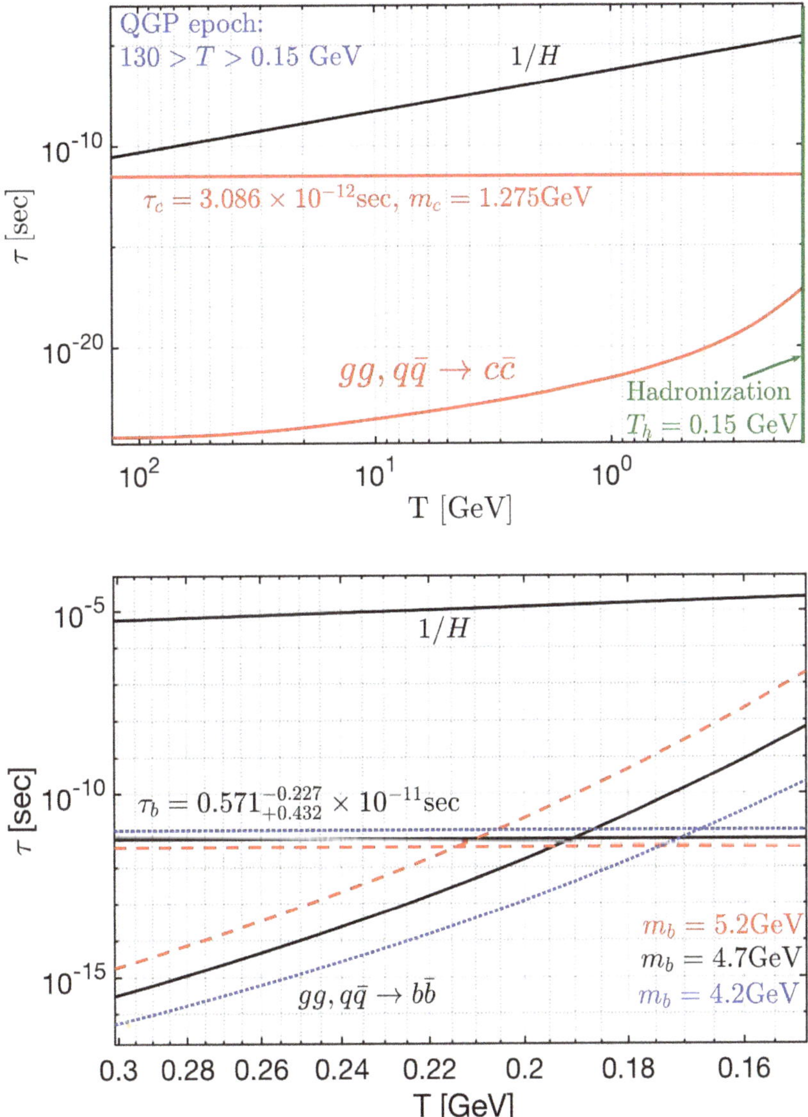

Figure 8. Comparison of Hubble time $1/H$, quark lifespan τ_q, and characteristic time for production via quark-gluon pair fusion for (**Top** figure) charm and (**Bottom** figure) bottom quarks as a function of temperature. Both figures end at approximately the hadronization temperature of $T_h \approx 150$ MeV. Three different masses $m_b = 4.2$ GeV (blue short dashes), 4.7 GeV, (solid black), 5.2 GeV (red long dashes) for bottom quarks are plotted to account for its decay width.

Let us describe the dynamical non-equilibrium of bottom quark abundance in QGP in more detail. The competition between decay and production reaction rates for bottom quarks in the early Universe can be written as

$$\frac{1}{V}\frac{dN_b}{dt} = \left(1 - Y_b^2\right) R_b^{\text{Source}} - Y_b R_b^{\text{Decay}} \,, \tag{24}$$

where N_b is the bottom quark abundance, Y_b is the general fugacity of bottom quarks, and R_b^{Source} and R_b^{Decay} are the thermal reaction rates per volume of production and decay of bottom quark, respectively [63]. The bottom source rate is controlled by quark-gluon pair fusion rate which vanishes upon hadronization. The decay rate depends on whether the bottom quarks are unconfined and free or bound within B-mesons which is controlled by the plasma temperature. Under the adiabatic approximation , we solve for the generalized bottom fugacity Y_b in Equation (24) yielding

$$Y_b = \frac{R_b^{\text{Decay}}}{2R_b^{\text{Source}}}\left[\sqrt{1 + \left(2R_b^{\text{Source}}/R_b^{\text{Decay}}\right)^2} - 1\right].\tag{25}$$

In Figure 9 we show the fugacity of the bottom quarks as a function of temperature $T = 0.3 \sim 0.15$ GeV for different masses of bottom quarks. In all cases, we have prolonged non-equilibrium $Y_b \neq 1$ because the decay and production rates of bottom quarks are of comparable temporal size to one another. The bottom content of QGP is exhausted as $Y_b \to 0$ as the Universe cools in temperature. For smaller masses, some bottom quark content is preserved up until hadronization as the strong interaction formation rate slows the depletion from weak decay near the QGP to HG phase transformation.

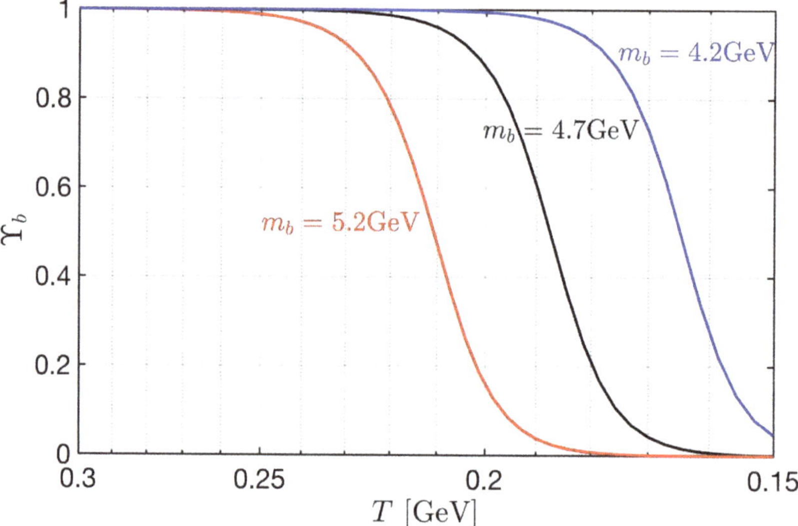

Figure 9. The generalized fugacity Y_b of free unconfined bottom quark as a function of temperature in QGP up to the hadronization temperature of $T_h \approx 150$ MeV for three different bottom masses $m_b = 4.2$ GeV (solid blue), 4.7 GeV, (solid black), 5.2 GeV (solid red).

As demonstrated above, the bottom quark flavor is capable to imprint arrow in time on physical processes being out of chemical equilibrium during the epoch $T = 0.3 \sim 0.15$ GeV. This is one of the required Sakharov condition (see Section 1.2) for baryogenesis. Our results provide a strong motivation to explore the physics of baryon non-conservation involving the bottom quarks and bound $b\bar{b}$ bottonium states in a thermal environment. Given that the non-equilibrium of bottom flavor arises at a relatively low QGP temperature allows for the baryogenesis to occur across primordial QGP hadronization epoch [63]. This result establishes the temperature era for the non-equilibrium abundance of bottom quarks.

3. Hadronic Epoch

3.1. The Formation of Matter

It is in this epoch that the matter of the Universe, including all the baryons which make up visible matter today, was created [61,62]. Unlike the fundamental particles, such as the quarks or W and Z, the mass of these hadrons is not due to the Higgs mechanism, but rather from the condensation of the QCD vacuum [13,64,65]. The quarks from which protons and neutrons are made have a mass more than 100 times smaller than these nucleons. The dominant matter mass-giving mechanism arises from quark confinement [66]. Light quarks are compressed by the quantum vacuum structure into a small space domain a hundred times smaller than their natural 'size'. A heuristic argument can be made by considering the variance in valance quark momentum Δp required by the Heisenberg uncertainty principle by confining them to a space of order $\Delta x \approx 1$ fm and the energy density of the attractive gluon field required to balance that outward pressure. That energy cost then manifests as the majority of the nucleon mass. The remaining few percent of mass is then due to the fact that quarks also have inertial mass provided by the Higgs mechanism as well as the electromagnetic mass for particles with charge.

The QGP-hadronization transformation is not instantaneous and involves a transitory period containing both hadrons and QGP [62]. Therefore the conservation laws outlined in Equations (15)–(17) can be violated in one phase as long as it is equally compensated in the other phase. This means the partition function during hadronization, and thus the formation of matter, should be parameterized between the hadron gas (HG) component and QGP component as

$$\ln \mathcal{Z}_{tot} = f_{HG}(T) \ln \mathcal{Z}_{HG} + [1 - f_{HG}(T)] \ln \mathcal{Z}_{QGP}, \tag{26}$$

where $f_{HG}(T)$ is the proportion of the phase space occupied by the hadron gas with values between $0 < f_{HG} < 1$. The charge neutrality condition Equation (15) is then modified to be

$$n_{Q,HG+QGP} = f_{HG}(T)n_{HG,Q} + [1 - f_{HG}(T)]n_{QGP,Q} = 0. \tag{27}$$

At a temperature of $T_h \approx 150$ MeV, the quarks and gluons become confined and condense into hadrons (both baryons and mesons). During this period, the number of baryon-antibaryon pairs is sufficiently high that the asymmetry (of ~ 1 in 10^9) would be essentially invisible until a temperature of between 40–50 MeV. We note that CPT symmetry is protected by the lack of asymmetry in normal Standard Model reactions to some large factor by the accumulation of scattering events through the majority of the Universe's evolution. CPT-violation is similarly restricted by possible mass difference in the Kaons [67] via the hypothetical difference in strange-antistrange quark masses which are expected to be small if not identically zero.

In Figure 10, we present the fraction of visible radiation and matter split between the baryons, mesons, and photons and leptons. For a brief early Universe period after QGP hadronization when the large amount of antimatter found in antiquarks converted into the dense gas of hadrons, their contribution to the energy density of the Universe competed with that of radiation and leptons [62]. Mass of matter will not emerge again until the late Universe after recombination though by that point dark matter would become the dominant form of matter in the cosmos.

The chemical potential of baryons after hadronization can be determined by the conserved baryon-per-entropy ratio under adiabatic expansion. Considering the net baryon density in the early Universe with temperature range $150 \, \text{MeV} > T > 5 \, \text{MeV}$ [24] we write

$$\frac{\left(n_B - n_{\overline{B}}\right)}{s} = \frac{1}{s}\left[(n_p - n_{\overline{p}}) + (n_n - n_{\overline{n}}) + (n_Y - n_{\overline{Y}})\right]$$

$$= \frac{45}{2\pi^4 g_*^s} \sinh\left[\frac{\mu_B}{T}\right] F_N \left[1 + \frac{F_Y}{F_N}\sqrt{\frac{1 + e^{-\mu_B/T} F_Y/F_K}{1 + e^{\mu_B/T} F_Y/F_K}}\right]. \tag{28}$$

where μ_B is the baryon chemical potential, g_*^s represents the effective entropic degrees of freedom, and we employ phase-space functions F_i for the set of nucleon N, kaon K, and hyperon Y particles. These functions are defined in Section 11.4 of [58] and given by

$$F_N = \sum_{N_i} g_{N_i} W(m_{N_i}/T), \quad N_i = n, p, \Delta(1232), \tag{29}$$

$$F_K = \sum_{K_i} g_{K_i} W(m_{K_i}/T), \quad K_i = K^0, \overline{K^0}, K^\pm, K^*(892), \tag{30}$$

$$F_Y = \sum_{Y_i} g_{Y_i} W(m_{Y_i}/T), \quad Y_i = \Lambda, \Sigma^0, \Sigma^\pm, \Sigma(1385), \tag{31}$$

where g_{N_i, K_i, Y_i} is the degeneracy of each baryonic species. We define the function $W(x) = x^2 K_2^B(x)$ where K_2^B is the modified Bessel functions of integer order "2".

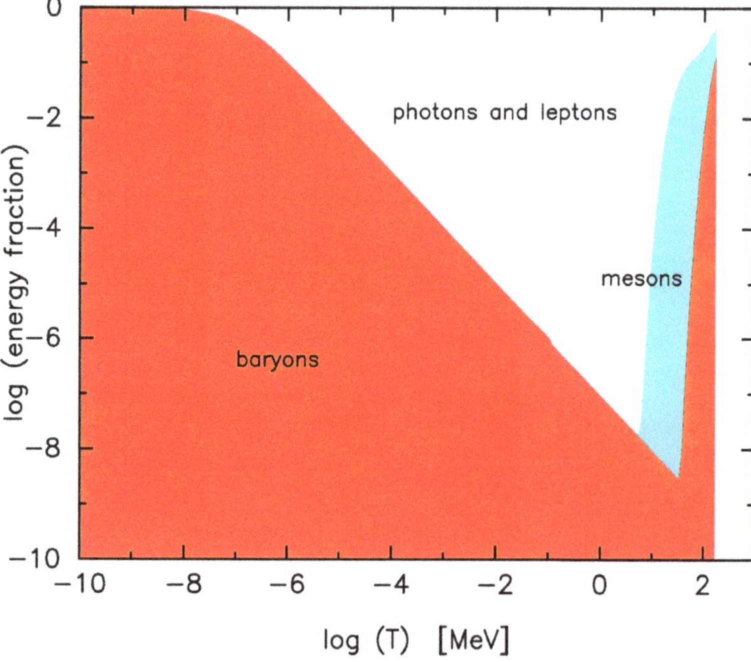

Figure 10. The fractional energy density of the luminous Universe (photons and leptons (white), mesons (blue), and hadrons (red)) as a function of the temperature of the Universe from hadronization to the contemporary era. This figure is a companion figure to Figure 2 (2003 unpublished, Fromerth & Rafelski [62]).

The net baryon-per-entropy-ratio can be obtained from the present-day measurement of the net baryon-per-photon ratio $(n_B - n_{\overline{B}})/n_\gamma$, where n_γ is the contemporary photon number density from the CMB [24]. This value is determined to be

$$\frac{n_B - n_{\overline{B}}}{s} = \left.\frac{n_B - n_{\overline{B}}}{s}\right|_{t_0} = (0.865 \pm 0.008) \times 10^{-10}. \tag{32}$$

We arrive at this ratio from considering the observed baryon-per-photon ratio [68] of

$$\frac{n_B - n_{\overline{B}}}{n_\gamma} = (0.609 \pm 0.006) \times 10^{-9}, \tag{33}$$

as well as the entropy-per-particle [23] for massless bosons and fermions

$$s/n|_{\text{boson}} \approx 3.60, \qquad s/n|_{\text{fermion}} \approx 4.20. \qquad (34)$$

Considering the inventory of strange mesons and baryons in the cosmos after hadronization, we evaluated the temperature of the net baryon disappearance in Figure 11. In solving Equation (28) numerically, we plot the baryon and antibaryon number density as a function of temperature in the range $150\,\text{MeV} > T > 5\,\text{MeV}$. The temperature where antibaryons disappear from the Universe inventory can be defined when the ratio $n_{\overline{B}}/(n_B - n_{\overline{B}}) = 1$. This condition was reached at temperature $T = 38.2$ MeV which is in agreement with the qualitative result in Kolb and Turner [69]. After this temperature, the net baryon density dilutes with a residual co-moving conserved quantity determined by the baryon asymmetry.

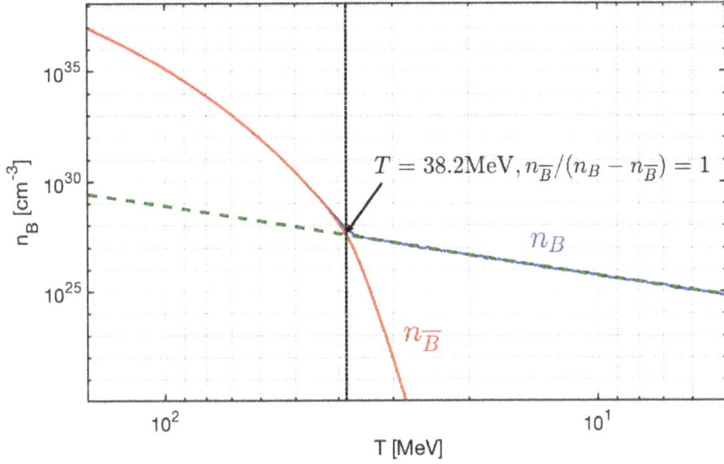

Figure 11. The baryon (blue solid line) and antibaryon (red solid line) number density as a function of temperature in the range $150\,\text{MeV} > T > 5\,\text{MeV}$. The green dashed line is the extrapolated value for baryon density. The temperature $T = 38.2$ MeV (black dashed vertical line) is denoted when the ratio $n_{\overline{B}}/(n_B - n_{\overline{B}}) = 1$ which define the condition where antibaryons disappear from the Universe.

The antibaryon disappearance temperature does not depend on baryon and lepton number neutrality $L = B$. Rather, it depends only on the baryon-per-entropy ratio which is assumed to be constant during the Universe's evolution, a condition which is maintained well after the plasmas discussed here vanish. The assumption of co-moving baryon number conservation is justified by the wealth of particle physics experiments, and the co-moving entropy conservation in an adiabatic evolving Universe is a common assumption.

3.2. Strangeness Abundance

As the energy contained in QGP is used up to create mesons, that is massive particles containing matter and antimatter, the high abundance of (anti)strange $(s, \bar{s})$ quark pairs present in the plasma is preserved. A smaller abundance of (anti)charm $(c, \bar{c})$ can combine with abundant strange quarks to form 'exotic' heavy mesons. With time, charmness and later strangeness decay away as these flavors are heavier than the light (u, d) quarks and antiquarks. Unlike charm, which disappears from the particle inventory relatively quickly, strangeness can still persist [24] in the Universe until $T \approx \mathcal{O}\,(10\,\text{MeV})$. As already noted, the meson sector is of particular interest in our work since mesons carry antimatter in form of their antiquark component. After the loss of antibaryons at $T = 38.2$ MeV, Figure 11, the remaining light mesons then act as a proxy for the hadronic antimatter evolution.

We illustrate this by considering an unstable strange particle S decaying into two particles 1 and 2 which themselves have no strangeness content. In a dense and high-temperature plasma with particles 1 and 2 in thermal equilibrium, the inverse reaction populates the system with particle S. This is written schematically as

$$S \Longleftrightarrow 1 + 2, \qquad \text{Example}: K^0 \Longleftrightarrow \pi + \pi. \tag{35}$$

The natural decay of the daughter particles provides the intrinsic strength of the inverse strangeness production reaction rate. As long as both decay and production reactions are possible, particle S abundance remains in thermal equilibrium. This balance between production and decay rates is called a detailed balance. The thermal reaction rate per time and volume for two-to-one particle reactions $1 + 2 \to 3$ has been presented before [70,71]. In full kinetic and chemical equilibrium, the reaction rate per time per volume is given by [71]:

$$R_{12 \to 3} = \frac{g_3}{(2\pi)^2} \frac{m_3}{\tau_3^0} \int_0^\infty \frac{p_3^2 dp_3}{E_3} \frac{e^{E_3/T}}{e^{E_3/T} \pm 1} \Phi(p_3) \,, \tag{36}$$

where τ_3^0 is the vacuum lifetime of particle 3. The positive sign "$+$" is for the case when particle 3 is a boson, while it is negative "$-$" for fermions. The function $\Phi(p_3)$ in the non-relativistic limit $m_3 \gg p_3, T$ can be written as

$$\Phi(p_3 \to 0) = 2 \frac{1}{(e^{E_1/T} \pm 1)(e^{E_2/T} \pm 1)}. \tag{37}$$

When back-reactions are faster than the Universe expansion, a condition we characterize in the following, we can explore the Universe composition assuming both kinetic and particle abundance equilibrium (chemical equilibrium). In Figure 12 we numerically solve for the chemical potential of strangeness and show the chemical equilibrium particle abundance ratios [24] for various mesons, the baryons, and their antiparticles. In the temperature range $150\,\mathrm{MeV} > T > 40\,\mathrm{MeV}$ the Universe is rich in physics phenomena involving strange mesons and (anti)baryons including (anti)hyperon abundances. While antibaryons vanish after temperature $T \approx 40\,\mathrm{MeV}$, kaons persist compared to baryons until $T = 20\,\mathrm{MeV}$. For temperatures $T < 20\,\mathrm{MeV}$, the Universe becomes light-quark baryons dominant. Pions $\pi(q\bar{q})$ persist the longest of the mesons (a feature explored in Section 3.3) until $T = 5.6\,\mathrm{MeV}$. Pions are the most abundant hadrons in this period because of their low mass and the inverse decay reaction $\gamma + \gamma \to \pi^0$ which assures chemical equilibrium [70].

Below $T = 5.6\,\mathrm{MeV}$, we have $n_\pi/n_B < 1$ and the number density of pion become sub-dominate compared to the remaining baryons. It is important to realize that hadrons always are a part of the evolving Universe, a point we wish to see emphasized more in literature. For temperatures $150\,\mathrm{MeV} > T > 20\,\mathrm{MeV}$ the Universe is meson-dominant with (anti)strangeness well represented in the meson sector with $s = \bar{s}$. Below temperature $T < 13\,\mathrm{MeV}$, strangeness inventory is mostly found in the hyperons as we have $(s - \bar{s}) \neq 0$. We note that hyperons never exceed baryon content throughout the hadron epoch. This period of meson physics ends the stage of the Universe where antimatter was dominant in the quark sector.

In Figure 13 we schematically show important source reactions for strange quark abundance in baryons and mesons considering both open and hidden strangeness ($s\bar{s}$-content). The important strangeness processes (involving both the quark and lepton sectors) are

$$\begin{aligned} l^- + l^+ &\leftrightarrow \phi, & \rho + \pi &\leftrightarrow \phi, \\ \pi + \pi &\leftrightarrow K, & \Lambda &\leftrightarrow \pi + N, & \mu^\pm + \nu &\leftrightarrow K^\pm. \end{aligned} \tag{38}$$

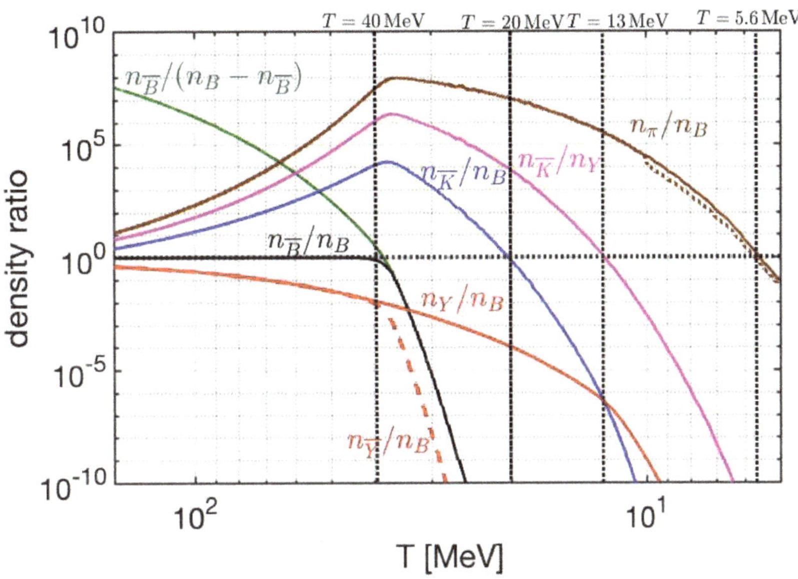

Figure 12. Ratios of hadronic particle number densities as a function of temperature 150 MeV > T > 5 MeV in the early Universe with baryon B yields: Pions $\pi(q\bar{q})$ (brown line), kaons $K(q\bar{s})$ (blue line), antibaryon $\overline{B}$ (black line), hyperon Y (red line) and antihyperons $\overline{Y}$ (dashed red line). Also shown is the $\overline{K}/Y$ ratio (purple line) and the $\bar{B}$ to asymmetry $B - \bar{B}$ ratio (green line). Temperature crossings are included (as vertical dashed black lines) at $T = 40$ MeV, 20 MeV, 13 MeV, 5.6 MeV as different abundances become sub-dominate compared to other species. The dashed brown line represents the drop in overall pion π abundance when the vanishing of the charged pions $\pi^{\pm}$ from the particle inventory is taken into account.

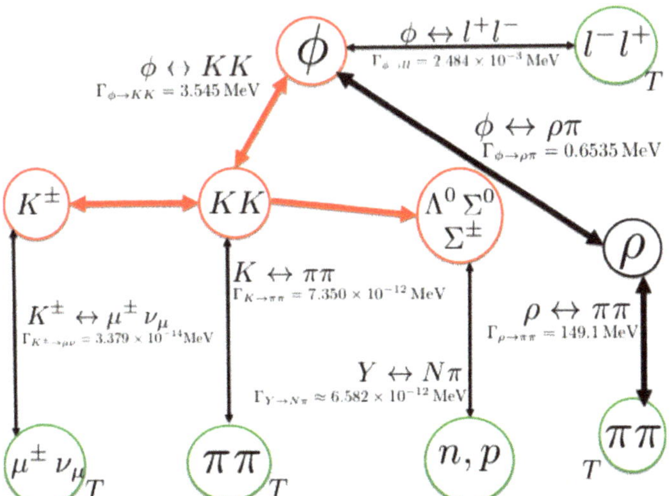

Figure 13. The strangeness abundance changing reactions in the primordial Universe. Red circles show strangeness carrying hadronic particles and thick red lines denote effectively instantaneous reactions. Thick black lines show relatively strong hadronic reactions.

Muons and pions are coupled through electromagnetic reactions

$$\mu^+ + \mu^- \leftrightarrow \gamma + \gamma, \qquad \pi^0 \leftrightarrow \gamma + \gamma, \tag{39}$$

to the photon background and retain their chemical equilibrium respectively [70,72]. The large $\phi \leftrightarrow K + K$ rate assures ϕ and K are in relative chemical equilibrium.

Once the primordial Universe expansion rate (given as the inverse of the Hubble parameter $1/H$) overwhelms the strongly temperature-dependent back-reaction, the decay $S \to 1 + 2$ occurs out of balance and particle S disappears from the Universe. In order to determine where exactly strangeness disappears from the Universe inventory we explore the magnitudes of a relatively large number of different rates of production and decay processes and compare these with the Hubble time constant [24]. Strangeness then primarily resides in two domains:

- Strangeness in the mesons
- Strangeness in the (anti)hyperons

In the meson domain, the relevant interaction rates competing with Hubble time are the reactions

$$\begin{aligned} \pi + \pi &\leftrightarrow K, \qquad \mu^\pm + \nu \leftrightarrow K^\pm, \\ l^+ + l^- &\leftrightarrow \phi, \qquad \rho + \pi \leftrightarrow \phi, \qquad \pi + \pi \leftrightarrow \rho. \end{aligned} \tag{40}$$

The relaxation times τ_i for these processes are compared with Hubble time in Figure 14. The criteria for a detailed reaction balance is broken once a process crosses above the Hubble time $1/H$ and thus can no longer be considered as subject to adiabatic evolution. As the Universe cools, these various processes freeze out as they cross this threshold. In Table 1 we show the characteristic strangeness reactions and their freeze-out temperatures in the hadronic epoch.

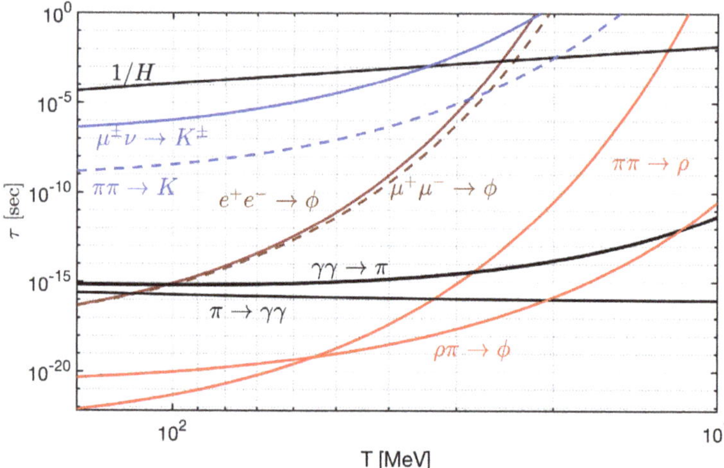

Figure 14. The hadronic reaction relaxation times τ_i in the meson sector as a function of temperature compared to Hubble time $1/H$ (black solid line). The following processes are presented: The leptonic (solid blue line) and strong (dashed blue line) kaon K processes, the electronic (solid dark red line) and muonic (dashed dark red line) phi meson ϕ processes, the forward and backward (thick black lines) electromagnetic pion π processes, and the strong (red lines) rho meson ρ processes.

Table 1. The characteristic strangeness reaction, their freeze-out temperature, and temperature width in the hadronic epoch.

Reactions	Freeze-Out Temperature (MeV)	ΔT_f (MeV)
$\mu^{\pm}\nu \to K^{\pm}$	$T_f = 33.8\,\text{MeV}$	3.5 MeV
$e^+e^- \to \phi$	$T_f = 24.9\,\text{MeV}$	0.6 MeV
$\mu^+\mu^- \to \phi$	$T_f = 23.5\,\text{MeV}$	0.6 MeV
$\pi\pi \to K$	$T_f = 19.8\,\text{MeV}$	1.2 MeV
$\pi\pi \to \rho$	$T_f = 12.3\,\text{MeV}$	0.2 MeV

Once freeze-out occurs and the corresponding detailed balance is broken, the inverse decay reactions act like a "hole" in the strangeness abundance siphoning strangeness out of the Universe's particle inventory. The first freeze-out reaction is the weak interaction kaon production process

$$\mu^{\pm} + \nu_{\mu} \to K^{\pm}, \qquad T_f^{K^{\pm}} = 33.8\,\text{MeV}, \tag{41}$$

which is followed by the electromagnetic ϕ meson production process

$$l^- + l^+ \to \phi, \qquad T_f^{\phi} = 23 \sim 25\,\text{MeV}. \tag{42}$$

Hadronic kaon production via pions follows next in the freeze-out process

$$\pi + \pi \to K, \qquad T_f^{K} = 19.8\,\text{MeV}. \tag{43}$$

as it becomes slower than the Hubble expansion. The reactions

$$\gamma + \gamma \leftrightarrow \pi, \qquad \rho + \pi \leftrightarrow \phi \tag{44}$$

remain faster compared to $1/H$ for the duration of the hadronic plasma epoch. Most ρ meson decays are faster [68] than ρ meson producing processes and cannot contribute to the strangeness creation in the meson sector. Below the temperature $T < 20$ MeV, all the detail balances in the strange meson sector are broken by freeze-out and the strangeness inventory in meson sector disappears rapidly.

Were it not for the small number of baryons present, strangeness would entirely vanish with the loss of the mesons. In order to understand strangeness in hyperons in the baryonic domain, we evaluated the reactions

$$\pi + N \leftrightarrow K + \Lambda, \qquad \overline{K} + N \leftrightarrow \Lambda + \pi, \qquad \Lambda \leftrightarrow N + \pi, \tag{45}$$

for strangeness production, exchange, and decay respectively in detail. The general form for thermal reaction rate per volume is discussed in Ch. 17 of [58]. In Figure 15 we show that for $T < 20$ MeV, the reactions for the hyperon Λ production is dominated by $\overline{K} + N \leftrightarrow \Lambda + \pi$. Both strangeness and antistrangeness disappear from the Universe via the reactions

$$\Lambda \to N + \pi, \qquad K \to \pi + \pi, \tag{46}$$

which conserves $s = \bar{s}$. Beginning with $T = 12.9$ MeV, the dominant reaction is $\Lambda \leftrightarrow N + \pi$, which shows that at lower temperatures strangeness content resides in the Λ baryon. This behavior is seen explicitly in Figure 12 where the hyperon abundance (of which the Λ baryon is a member) exceeds the rapidly diminishing kaon abundance as the Universe cools. While hyperons never form a dominant component of the hadronic content of the Universe, it is an important life-boat for strangeness persisting after the more transitory mesons. In this case, the strangeness abundance becomes asymmetric and we have $s \gg \bar{s}$ at temperatures $T < 12.9$ MeV. Hence, strange hyperons and antihyperons could enter into dynamic non-equilibrium condition including $\langle s - \bar{s} \rangle \neq 0$. The primary conclusion of the study of strangeness production and content in the early Universe, following on QGP hadronization, is that the relevant temperature domains indicate a complex interplay between baryon and meson (strange and non-strange) abundances and non-trivial decoupling from equilibrium for strange and non-strange mesons.

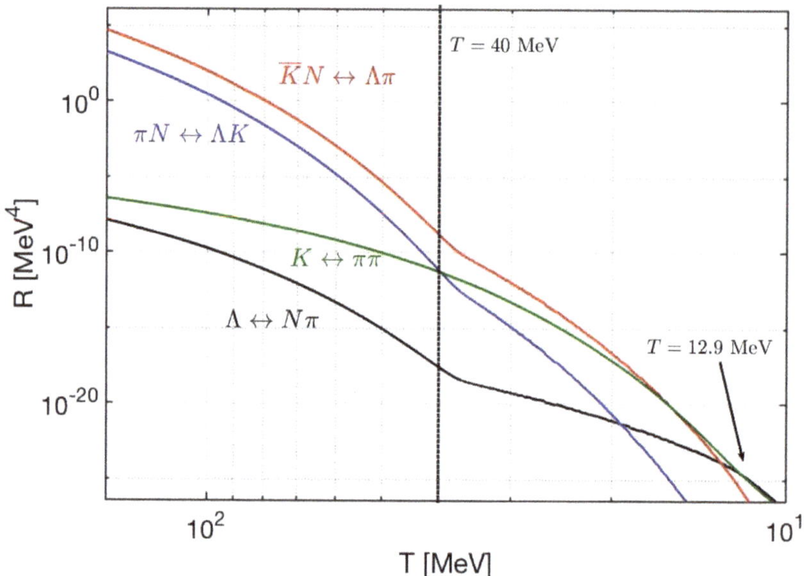

Figure 15. Thermal reaction rate R per volume and time for important hadronic strangeness production, exchange and decay processes as a function of temperature $150\,\text{MeV} > T > 10\,\text{MeV}$. The following processes are presented: $\Lambda \leftrightarrow N\pi$ (solid black line), $K \leftrightarrow \pi\pi$ (solid green line), $\pi N \leftrightarrow \Lambda K$ (solid blue line), $\bar{K}N \leftrightarrow \Lambda\pi$ (solid red line). Two temperature crossings are denoted at $T = 40$ MeV, 12.9 MeV.

3.3. Pion Abundance

Pions ($q\bar{q}, q \in u, d$), the lightest hadrons, are the dominant hadrons in the hadronic era and the most abundant hadron family well into the leptonic epoch (see Section 4). The neutral pion π^0 vacuum lifespan of $\tau_{\pi^0}^0 = (8.52 \pm 0.18) \times 10^{-17}$ s [68] is far shorter compared to the Hubble expansion time of $1/H = (10^{-3} \sim 10^{-4})$ s within this epoch as depicted in Figure 14.

At seeing such a large discrepancy in characteristic times, one is tempted to presume that the decay process dominates and that π^0 disappears quickly in the hadronic gas. However, in the high temperature $T = \mathcal{O}\,(100\,\text{MeV}) \sim \mathcal{O}\,(10\,\text{MeV})$ thermal bath of this era, the inverse decay reaction forms neutral pions π^0 at rate corresponding to the decay process maintaining the abundance of the species (see Figure 12). In general, π^0 is produced in the QED plasma predominantly by thermal two-photon fusion:

$$\gamma + \gamma \rightarrow \pi^0. \tag{47}$$

This formation process is simply the inverse of the dominant decay process. While we do not address it in detail here, the $\pi^{\pm}$ charged pions are also in thermal equilibrium with the other pions species via hadronic and electromagnetic reactions

$$\pi^0 + \pi^0 \leftrightarrow \pi^+ + \pi^- \qquad l^+ + l^- \leftrightarrow \pi^+ + \pi^-, \qquad \gamma + \gamma \leftrightarrow \pi^+ + \pi^-. \tag{48}$$

Of these, the hadronic interaction is the fastest and controls the charged pion abundance most directly [23,73] such that the condition

$$\rho_{\pi^0} \sim \rho_{\pi^\pm},\tag{49}$$

where ρ is the energy density of the species and is maintained for most of the hadronic era. We point out that the in the late (colder) hadronic era, the charged pions will scatter off the remaining baryons with asymmetric reactions due to the lack of antibaryons. The smallness of the electronic e^+e^- formation of π^0 is characterized by its small branching ratio in π^0 decay $B = \Gamma_{ee}/\Gamma_{\gamma\gamma} = 6.2 \pm 0.5 \times 10^{-8}$ [68] which can be neglected compared to photon fusion. The general form for invariant production rates and relaxation time is discussed in [70] where we have for the photon fusion process

$$R_{\gamma\gamma\to\pi^0} = \int \frac{d^3p_\pi}{(2\pi)^3 2E_\pi} \int \frac{d^3p_{2\gamma}}{(2\pi)^3 2E_{2\gamma}} \int \frac{d^3p_{1\gamma}}{(2\pi)^3 2E_{1\gamma}} (2\pi)^4 \delta^4\left(p_{1\gamma} + p_{2\gamma} - p_\pi\right) \times$$
$$\sum_{spin} |\langle p_{1\gamma}p_{2\gamma}|M|p_\pi\rangle|^2 f_\pi(p_\pi) f_\gamma(p_{1\gamma}) f_\gamma(p_{2\gamma}) Y_\gamma^{-2} Y_{\pi^0}^{-1} e^{u \cdot p_\pi/T},\tag{50}$$

where Y_i is the fugacity and f_i is the Bose-Einstein distribution of particle i, and M is the matrix element for the process. Since the $\gamma + \gamma \to \pi^0$ is the dominant mechanism of pion production, we can omit all sub-dominant processes, and the dynamic equation of π^0 abundance can be written as [23]:

$$\frac{d}{dt}Y_{\pi^0} = \frac{1}{\tau_T}Y_{\pi^0} + \frac{1}{\tau_S}Y_{\pi^0} + \frac{1}{\tau_{\pi^0}}\left(Y_\gamma^2 - Y_{\pi^0}\right),\tag{51}$$

where τ_T and τ_S are the kinematic relaxation times for temperature and entropy evolution and τ_{π^0} is the chemical relaxation time for π^0. We have

$$\frac{1}{\tau_T} \equiv -T^3 g^* \frac{d(n_\pi/(Y_3 g^* T^3))/dT}{dn_\pi/dY_3}\dot{T},$$
$$\frac{1}{\tau_S} \equiv -\frac{n_\pi/Y_3}{dn_\pi/dY_3}\frac{d\ln(g^* V T^3)}{dT}\dot{T},\tag{52}$$
$$\tau_{\pi^0} = \frac{dn_{\pi^0}/dY_{\pi^0}}{R_{\pi^0}},$$

where n_{π^0} is the number density of pions. A minus sign is introduced in the above expressions to maintain $\tau_T, \tau_S > 0$. Since entropy is conserved within the radiation-dominated epoch, we have $T^3 V = $ constant thus $d(T^3 V(T))/dT = 0$. This implies the entropic relaxation time is infinite yielding $1/\tau_S = 0$. The effect of Universe expansion and dilution of number density is described by $1/\tau_T$. Comparing τ_T to the chemical relaxation time τ_{π^0} can provide the quantitative condition for freeze-out from chemical equilibrium. In the case of pion mass being much larger than the temperature, $m_\pi \gg T$, we have [73]

$$\tau_T \approx \frac{T}{m_\pi H}.\tag{53}$$

In Figure 14 we compare the relaxation time of τ_{π^0} to the Hubble time $1/H$ which shows that $\tau_{\pi^0} \ll 1/H$. In such a case, the yield of π^0 is expected to remain in chemical equilibrium (even as its thermal number density gradually decreases) with no freeze-out temperature occurring. This makes pions distinct from all other meson species. This phenomenon can be attributed to the high population of photons as in such an environment, it remains sufficiently probable to find high-energy photons to fuse back into neutral pions π^0 [23] for the duration of large pion abundance. As shown in Figure 12, pions remained as proxy for hadronic matter and antimatter down to $T = 5.6$ MeV.

4. Leptonic Epoch

4.1. Thermal Degrees of Freedom

The leptonic epoch, dominated by photons and both charged and neutral leptons, is notable for being the last time where neutrinos played an active role in the Universe's thermal dynamics before decoupling and becoming free-streaming. In the early stage of this plasma after the hadronization era ended $T \approx \mathcal{O}\,(10\ \mathrm{MeV})$, neutrinos represented the highest energy density followed by the light charged leptons and then finally the photons. The differing relativistic limit energy densities can be related by

$$\rho_{e^\pm} \approx \left(2 \times \frac{7}{8}\right)\rho_\gamma, \qquad \rho_\nu \approx \left(3 \times \frac{7}{8}\right)\rho_\gamma. \tag{54}$$

The reason for this hierarchy is because of the degrees of freedom [29,58] available in each species in thermal equilibrium; the factor $7/8$ arises from the difference in pressure contribution between bosons and fermions.

While photons only exhibit two polarization degrees of freedom, the charged light leptons could manifest as both matter (electrons), antimatter (positrons) and as well as two polarizations yielding $2 \times 2 = 4$. The neutral leptons made up of the neutrinos however had three thermally active species $3 \times 2 = 6$ boosting their energy density in that period to more than any other contribution. The muon-antimuon energy density was also controlled by its degrees of freedom matching that of $e^\pm$ until $T \approx \mathcal{O}\,(100\ \mathrm{MeV})$, still well within the hadronic epoch, when the heavier lepton no longer satisfied the ultra-relativistic (and thus massless) limit. This separation of the two lighter charge lepton dynamics is seen in Figure 2 after hadronization.

The known cosmic degrees of freedom require that if and when neutrinos are Dirac-like and have chiral right-handed (matter) components, then these right handed components must not drive the neutrino effective degrees of freedom N_{eff}^ν away from three. In a more general context the non-interacting sterile neutrinos could also inflate N_{eff}^ν during this epoch for the same reasoning [74–78] or have a connection to dark matter [79,80]. The neutrino degrees of freedom will be more fully discussed in Section 4.5.

4.2. Muon Abundance

As seen in Section 3.2, muon abundance and their associated reactions are integral to the understanding of the strangeness and antistrangeness content of the primordial Universe [24]. Therefore we determine to what extent and temperature (anti)muons remained in chemical abundance equilibrium. Without a clear boundary separating the hadronic epoch from the leptonic epoch, there is complete overlap in the hadronic and leptonic species dynamics in the period $T = \mathcal{O}\,(10\ \mathrm{MeV}) \sim \mathcal{O}\,(1\ \mathrm{MeV})$.

In the cosmic plasma, muons can be produced by predominately electromagnetic and weak interaction processes

$$\gamma + \gamma \longrightarrow \mu^+ + \mu^-, \qquad\qquad e^+ + e^- \longrightarrow \mu^+ + \mu^-, \tag{55}$$

$$\pi^- \longrightarrow \mu^- + \bar{\nu}_\mu, \qquad\qquad \pi^+ \longrightarrow \mu^+ + \nu_\mu. \tag{56}$$

Provided that all particles shown on the left-hand side of each reaction (namely the photons, electrons (positrons) and charged pions) exist in chemical equilibrium, the back-reaction for each of the above processes occurs in detailed balance.

The scattering angle averaged thermal reaction rate per volume for the reaction $a\bar{a} \to b\bar{b}$ in Boltzmann approximation is given by [58]

$$R_{a\bar{a}\to b\bar{b}} = \frac{g_a g_{\bar{a}}}{1+I}\frac{T}{32\pi^4}\int_{s_{th}}^\infty ds\,\frac{s(s-4m_a^2)}{\sqrt{s}}\sigma_{a\bar{a}\to b\bar{b}}K_1(\sqrt{s}/T), \tag{57}$$

where s_{th} is the threshold energy for the reaction, $\sigma_{a\bar{a}\to b\bar{b}}$ is the cross section for the given re-action. We introduce the factor $1/(1+I)$ to avoid the double counting of indistinguishable pairs of particles where $I=1$ for an identical pair and $I=0$ for a distinguishable pair.

The muon weak decay processes are

$$\mu^- \to \nu_\mu + e^- + \bar{\nu}_e, \qquad \mu^+ \to \bar{\nu}_\mu + e^+ + \nu_e, \tag{58}$$

with the vacuum life time $\tau_\mu = 2.197 \times 10^{-6}$ s producing (anti)neutrino pairs of differing flavor and electrons(positrons). We recall the considerable shorter vacuum lifetime of pions $\tau_{\pi^\pm} = 2.6033 \times 10^{-8}$ s. The thermal decay rate per volume in the Boltzmann limit is [70]

$$R_i = \frac{g_i}{2\pi^2}\left(\frac{T^3}{\tau_i}\right)\left(\frac{m_i}{T}\right)^2 K_1(m_i/T) \tag{59}$$

where τ_i is the vacuum lifespan of a given particle i.

These production and decay rates for muonic processes are evaluated in [72]. From this, we can determine the temperature when muons rather suddenly disappear from the particle inventory of the Universe which occurs when their decay rate exceeds their production rate. In Figure 16 we show the invariant thermal reaction rates per volume and time for the relevant muon reactions. As the temperature decreases in the expanding Universe, the initially dominant production rates become rapidly smaller due to the mass threshold effect. This is allowing the production and decay rates to become equal. The characteristic times are much faster than the Hubble time (not shown in Figure 16). Muon abundance therefore disappears just when the decay rate overwhelms production at the temperature $T_{\rm dis} = 4.20$ MeV.

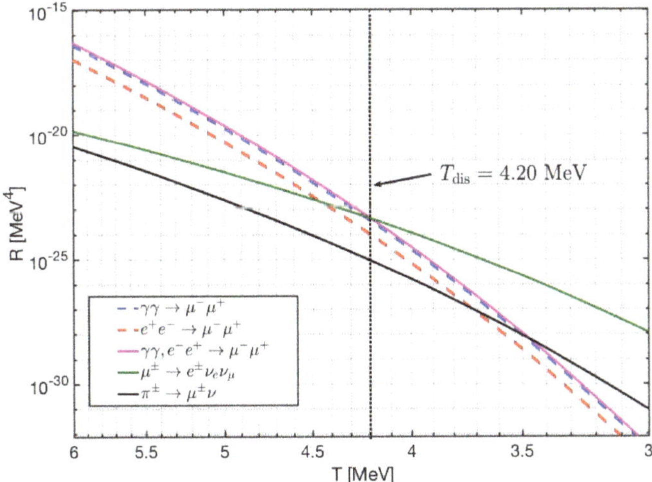

Figure 16. The thermal reaction rate per volume for muon related reactions as a function of tempera-ture adapted from [72]. The dominant reaction rates for $\mu^\pm$ production are: The $\gamma\gamma$ channel (blue dashed line), $e^\pm$ (red dashed line), these two combined as the total electromagnetic rate (pink solid line), and the charged pion decay feed channel (black solid line). The muon decay rate is also shown (green solid line). The crossing point between the electromagnetic production processes and the muonic decay rate is seen as the dashed vertical black line at $T_{\rm dis} = 4.2$ MeV.

In Figure 17 we show that the number density ratio of muons to baryons $n_{\mu^\pm}/n_B$ at the muon disappearance temperature $T_{\rm dis} = 4.20$ MeV is $n_{\mu^\pm}/n_B \approx 0.91$ [24]. Interestingly, this means that the muon abundance may still be able to influence baryon evolution up to this point because their number density is comparable to that of baryons (there are no antibaryons). This coincidence of abundance offers a novel and tantalizing model-building opportunity for both baryon-antibaryon separation models and/or strangelet formation models.

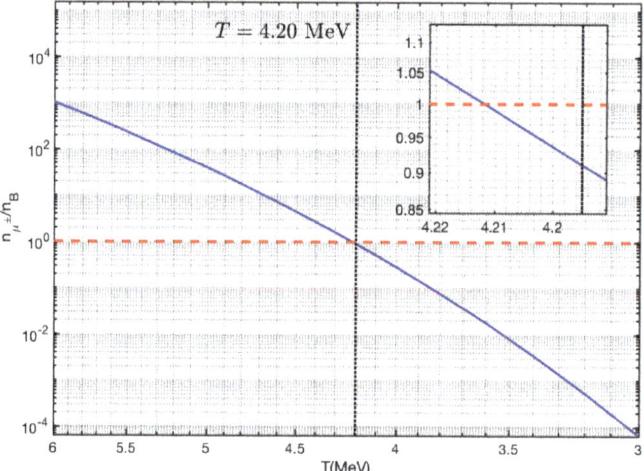

Figure 17. The density ratio between $\mu^\pm$ and baryons $n_{\mu^\pm}/n_B$ (blue solid line) is plotted as a function of temperature. The red dashed line indicates a density ratio value of $n_{\mu^\pm}/n_B = 1$. The density ratio at the muon disappearance temperature (vertical black dashed line) is about $n_{\mu^\pm}/n_B(T_{\mathrm{dis}}) \approx 0.911$.

4.3. Neutrino Masses and Oscillation

Neutrinos are believed to have a small, but nonzero mass due to the phenomenon of flavor oscillation [81–83]. This is seen in the flux of neutrinos from the Sun, and also in terrestrial reactor experiments. In the Standard Model neutrinos are produced via weak charged current (mediated by the W boson) as flavor eigenstates. If the neutrino was truly massless, then whatever flavor was produced would be immutable as the propagating state. However, if neutrinos have mass, then they propagate through space as their mass-momentum eigenstates. Neutrino masses can be written in terms of an effective theory where the mass term contains various couplings between neutrino states determined by some BSM theory. The exact form of such a BSM theory is outside the scope of this work, we refer the reader to some standard references [84–87].

Within the Standard Model keeping two degrees of freedom for each neutrino flavor the Majorana fermion mass term is given by

$$\mathcal{L}_m^{Maj.} = -\frac{1}{2}\bar{\nu}_L^\alpha M_{\alpha\beta}^M (\nu_L^\beta)^c + \text{h.c.}, \tag{60}$$

where $\nu^c = \hat{C}(\bar{\nu})^T$ is the charge conjugate of the neutrino field. The operator $\hat{C} = i\gamma^2\gamma^0$ is the charge conjugation operator. An interesting consequence of neutrinos being Majorana particles is that they would be their own antiparticles like photons allowing for violations of total lepton number. Neutrinoless double beta decay is an important, yet undetected, evidence for Majorana nature of neutrinos [88]. Majorana neutrinos with small masses can be generated from some high scale via the See-Saw mechanism [89–91] which ensures that the degrees of freedom separate into heavy neutrinos and light nearly massless Majorana neutrinos. The See-Saw mechanism then provides an explanation for the smallness of the neutrino masses as has been experimentally observed.

A flavor eigenstate ν^α can be described as a superposition of mass eigenstates ν^k with coefficients given by the Pontecorvo-Maki-Nakagawa-Sakata (PMNS) mixing matrix [92,93] which are both in general complex and unitary. This is given by

$$\nu^\alpha = \sum_k^n U^*_{\alpha k} \nu^k, \qquad \alpha = e, \mu, \tau, \qquad k = 1, 2, 3, \dots, n \tag{61}$$

where U is the PMNS mixing matrix. The PMNS matrix is the lepton equivalent to the CKM mixing matrix which describes the misalignment between the quark flavors and their masses. For Majorana neutrinos, there can be up to three complex phases (δ, ρ, γ) which are CP-violating [94] which are present when the number of generations is $n \geq 3$. For Dirac-like neutrinos, only the δ complex phase is required. In principle, the number of mass eigenstates can exceed three, but is restricted to three generations in most models. By standard convention [95] found in the literature we parameterize the rotation matrix U as

$$U = \begin{pmatrix} c_{12}c_{13} & s_{12}c_{13} & s_{13}e^{-i\delta} \\ -s_{12}c_{23} - c_{12}s_{13}s_{23}e^{i\delta} & c_{12}c_{23} - s_{12}s_{13}s_{23}e^{i\delta} & c_{13}s_{23} \\ s_{12}s_{23} - c_{12}s_{13}c_{23}e^{i\delta} & -c_{12}s_{23} - s_{12}s_{13}c_{23}e^{i\delta} & c_{13}c_{23} \end{pmatrix} \times \begin{pmatrix} 1 & & \\ & e^{i\rho} & \\ & & e^{i\gamma} \end{pmatrix}, \tag{62}$$

where $c_{ij} = \cos(\theta_{ij})$ and $s_{ij} = \sin(\theta_{ij})$. In this convention, the three mixing angles $(\theta_{12}, \theta_{13}, \theta_{23})$, are understood to be the Euler angles for generalized rotations.

The neutrino proper masses are generally considered to be small with values no more than 0.1 eV. Because of this, neutrinos produced during fusion within the Sun or radioactive fission in terrestrial reactors on Earth propagate relativistically. Evaluating freely propagating plane waves in the relativistic limit yields the vacuum oscillation probability between flavors ν_α and ν_β written as [96]

$$P_{\alpha \to \beta} = \delta_{\alpha\beta} - 4 \sum_{i<j}^n \mathrm{Re}\left[U_{\alpha i} U^*_{\beta i} U^*_{\alpha j} U_{\beta j}\right] \sin^2\left(\frac{\Delta m_{ij}^2 L}{4E}\right)$$
$$+ 2 \sum_{i<j}^n \mathrm{Im}\left[U_{\alpha i} U^*_{\beta i} U^*_{\alpha j} U_{\beta j}\right] \sin\left(\frac{\Delta m_{ij}^2 L}{2E}\right), \qquad \Delta m_{ij}^2 \equiv m_i^2 - m_j^2 \tag{63}$$

where L is the distance traveled by the neutrino between production and detection. The square mass difference Δm_{ij}^2 has been experimentally measured [96]. As oscillation only restricts the differences in mass squares, the precise values of the masses cannot be determined from oscillation experiments alone. It is also unknown under what hierarchical scheme (normal or inverted) [97,98] the masses are organized as two of the three neutrino proper masses are close together in value.

It is important to point out that oscillation does not represent any physical interaction (except when neutrinos must travel through matter which modulates the ν_e flavor [99,100]) or change in the neutrino during propagation. Rather, for a given production energy, the superposition of mass eigenstates each have unique momentum and thus unique group velocities. This mismatch in the wave propagation leads to the oscillatory probability of flavor detection as a function of distance.

We further note that non-interacting BSM so called sterile neutrinos of any mass have not yet been observed despite extensive searching. The existence of such neutrinos, if they were ever thermally active in the early cosmos would leave fingerprints on the Cosmic Neutrino Background (CNB) spectrum [78]. The presence of an abnormally large anomalous magnetic moment [84,101–106] for the neutrino would also possibly leave traces in the evolution of the early Universe.

4.4. Neutrino Freeze-Out

The relic neutrino background (or CNB) is believed to be a well-preserved probe of a Universe only a second old which at some future time may become experimentally accessible. The properties of the neutrino background are influenced by the details of the freeze-out or decoupling process at a temperature $T = \mathcal{O}\,(2\,\mathrm{MeV})$. The freeze-out process, whereby a particle species stops interacting and decouples from the photon background, involves several steps that lead to the species being described by the free-streaming momentum distribution. We outline freeze-out properties, including what distinguishes it from the equilibrium distributions [25].

Chemical freeze-out of a particle species occurs at the temperature, T_{ch}, when particle number changing processes slow down and the particle abundance can no longer be maintained at an equilibrium level. Prior to the chemical freeze-out temperature, number changing processes are significant and keep the particle in chemical (and thermal) equilibrium, implying that the distribution function has the Fermi-Dirac form, obtained by maximizing entropy at fixed energy (parameter $1/T$) and particle number (parameter λ)

$$f_c(t, E) = \frac{1}{\lambda \exp(E/T) + 1}, \ \text{for } T(t) > T_{ch}. \tag{64}$$

Kinetic freeze-out occurs at the temperature, T_f, when momentum exchanging interactions no longer occur rapidly enough to maintain an equilibrium momentum distribution. When $T_f < T(t) < T_{ch}$, the number-changing process no longer occurs rapidly enough to keep the distribution in chemical equilibrium but there is still sufficient momentum exchange to keep the distribution in thermal equilibrium. The distribution function is therefore obtained by maximizing entropy, with fixed energy, particle number, and antiparticle number separately. This implies that the distribution function has the form

$$f_k(t, E) = \frac{1}{\Upsilon^{-1} \exp(E/T) + 1}, \ \text{for } T_f < T(t) < T_{ch}. \tag{65}$$

The time dependent generalized fugacity $\Upsilon(t)$ controls the occupancy of phase space and is necessary once $T(t) < T_{ch}$ in order to conserve particle number.

For $T(t) < T_f$ there are no longer any significant interactions that couple the particle species of interest and so they begin to free-stream through the Universe, i.e., travel on geodesics without scattering. The Einstein-Vlasov equation can be solved, see [107], to yield the free-streaming momentum distribution

$$f(t, E) = \frac{1}{\Upsilon^{-1} e^{\sqrt{p^2/T^2 + m^2/T_f^2}} + 1} \tag{66}$$

where the free-streaming effective temperature

$$T(t) = \frac{T_f a(t_k)}{a(t)} \tag{67}$$

is obtained by redshifting the temperature at kinetic freeze-out. The corresponding free-streaming energy density, pressure, and number densities are given by

$$\rho = \frac{d}{2\pi^2} \int_0^\infty \frac{\left(m^2 + p^2\right)^{1/2} p^2 dp}{\mathrm{Y}^{-1} e^{\sqrt{p^2/T^2 + m^2/T_f^2}} + 1}, \tag{68}$$

$$P = \frac{d}{6\pi^2} \int_0^\infty \frac{\left(m^2 + p^2\right)^{-1/2} p^4 dp}{\mathrm{Y}^{-1} e^{\sqrt{p^2/T^2 + m^2/T_f^2}} + 1}, \tag{69}$$

$$n = \frac{d}{2\pi^2} \int_0^\infty \frac{p^2 dp}{\mathrm{Y}^{-1} e^{\sqrt{p^2/T^2 + m^2/T_f^2}} + 1}, \tag{70}$$

where d is the degeneracy of the particle species. These differ from the corresponding expressions for an equilibrium distribution in Minkowski space by the replacement $m \to mT(t)/T_f$ only in the exponential.

The separation of the freeze-out process into these three regimes is of course only an approximation. In principle, there is a smooth transition between them. However, it is a very useful approximation in cosmology. See [38,108] for methods capable of resolving these smooth transitions.

To estimate the freeze-out temperature we need to solve the Boltzmann equation with different types of collision terms. In [109] we detail a new method for analytically simplifying the collision integrals and show that the neutrino freeze-out temperature is controlled by standard model (SM) parameters. The freeze-out temperature depends only on the magnitude of the Weinberg angle in the form $\sin^2 \theta_W$, and a dimensionless relative interaction strength parameter η,

$$\eta \equiv M_p m_e^3 G_F^2, \qquad M_p^2 \equiv \frac{1}{8\pi G_N}, \tag{71}$$

a combination of the electron mass m_e, Newton constant G_N (expressed above in terms of Planck mass M_p), and the Fermi constant G_F. The dimensionless interaction strength parameter η in the present-day vacuum has the value

$$\eta_0 \equiv M_p m_e^3 G_F^2 \big|_0 = 0.04421. \tag{72}$$

The magnitude of $\sin^2 \theta_W$ is not fixed within the SM and could be subject to variation as a function of time or temperature. In Figure 18 we show the dependence of neutrino freeze-out temperatures for ν_e and $\nu_{\mu,\tau}$ on SM model parameters $\sin^2 \theta_W$ and η in detail. The impact of SM parameter values on neutrino freeze-out and the discussion of the implications and connections of this work to other areas of physics, namely Big Bang Nucleosynthesis and dark radiation can be found in detail in [109–112].

After neutrinos freeze-out, the neutrino co-moving entropy is independently conserved. However, the presence of electron-positron rich plasma until $T = 20$ keV provides the reaction $\gamma\gamma \to e^- e^+ \to \nu\bar{\nu}$ to occur even after neutrinos decouple from the cosmic plasma. This suggests the small amount of $e^\pm$ entropy can still transfer to neutrinos until temperature $T = 20$ keV and can modify free streaming distribution and the effective number of neutrinos.

We expect that incorporating oscillations into the freeze-out calculation would yield a smaller freeze-out temperature difference between neutrino flavors as oscillation provides a mechanism in which the heavier flavors remain thermally active despite their direct production becoming suppressed. In work by Mangano et al. [38], neutrino freeze-out including flavor oscillations is shown to be a negligible effect.

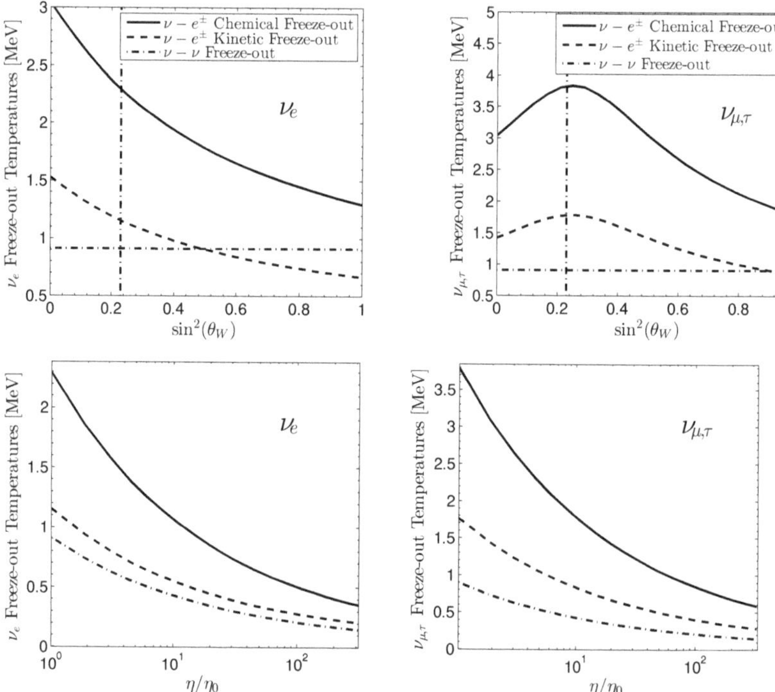

Figure 18. Freeze-out temperatures for electron neutrinos (**left**) and μ, τ neutrinos (**right**) for the three types of freeze-out processes adapted from paper [109]. Top panels print temperature curves as a function of $\sin^2 \theta_W$ for $\eta = \eta_0$, the vertical dashed line is $\sin^2 \theta_W = 0.23$; bottom panels are printed as a function of relative change in interaction strength η/η_0 obtained for $\sin^2 \theta_W = 0.23$.

4.5. Effective Number of Neutrinos

The population of each flavor of neutrino is not a fixed quantity throughout the evolution of the Universe. In the earlier hot Universe, the population of neutrinos is controlled thermally and to maximize entropy, each flavor is equally filled. As the expansion factor $a(t)$ is radiation dominated for much of this period (see Figure 2), the CMB is ultimately sensitive to the total energy density within the neutrino sector (which is sometimes referred to as the dark radiation contribution). This is described by the effective number of neutrinos N_ν^{eff} which captures the number of relativistic degrees of freedom for neutrinos as well as any reheating that occurred in the sector after freeze-out. This quantity is related to the total energy density in the neutrino sector as well as the photon background temperature of the Universe via

$$N_\nu^{\text{eff}} \equiv \frac{\rho_\nu^{\text{tot}}}{\frac{7\pi^2}{120}\left(\frac{4}{11}\right)^{4/3} T_\gamma^4}, \tag{73}$$

where ρ_ν^{tot} is the total energy density in neutrinos and T_γ is the photon temperature.

N_ν^{eff} is defined such that three neutrino flavors with zero participation of neutrinos in reheating during $e^\pm$ annihilation results in $N_\nu^{\text{eff}} = 3$. The factor of $(4/11)^{1/3}$ relates the photon temperature to the (effective) temperature of the free-streaming neutrinos after $e^\pm$ annihilation, under the assumption of zero neutrino reheating. Strictly speaking, the number of true degrees of freedom is exactly determined by the number of neutrino families and available quantum numbers, therefore deviations of $N_\nu^{\text{eff}} > 3$ are to be understood as reheating which goes into the neutrino energy density ρ_ν^{tot}.

Experimentally, N_{eff} has been determined from CMB data by the Planck collaboration [28] in their 2018 analysis yielding $N_{\nu,exp}^{\text{eff}} = 2.99 \pm 0.17$ though this value has evolved substantially since their 2013 and 2015 analyses [41,42]. Precise study of neutrino decoupling (as outlined in Section 4.4) and thus freeze-out can improve the predictions for the value of N_{ν}^{eff}. Many studies focus on improving the calculation of decoupling through various means such as

1. Determining the dependence of freeze-out on the natural constants found in the Standard Model of particle physics [26,109].
2. The entropy transfer from electron-positron annihilation and finite temperature correction at neutrino decoupling [39,113,114].
3. Neutrino decoupling with flavor oscillations [38,40]. Nonstandard neutrino interactions have been investigated, including neutrino electromagnetic [101–105,115] and nonstandard neutrino electron coupling [115].

As N_{ν}^{eff} is only a measure of the relativistic energy density leading up to photon decoupling, a natural alternative mechanism for obtaining $N_{\nu}^{\text{eff}} > 3$ is the introduction of additional, presently not discovered, weakly interacting massless particles [80,116–119]. Alternatively, theories outside conventional freeze-out considerations have been proposed to explain the tension in N_{eff} including: QGP as the possible source of N_{eff} or connection between lepton asymmetry L and N_{ν}^{eff}.

The natural consistency of the reported CMB range of N_{ν}^{eff} with the range of QGP hadronization temperatures, motivates the exploration of a connection between N_{ν}^{eff} and the decoupling of sterile particles at and below the QGP phase transition [120]. This demonstrates that that $N_{\nu}^{\text{eff}} > 3.05$ can be associated with the appearance of several light particles at QGP hadronization in the early Universe that either are weakly interacting in the entire space or is only allowed to interact within the deconfined domain, in which case their coupling would be strong. Such particles could leave a clear dark radiation experimental signature in relativistic heavy-ion experiments that produce the deconfined QGP phase.

In standard ΛCDM, the asymmetry between leptons and antileptons $L \equiv [N_{\text{L}} - N_{\bar{\text{L}}}]/N_{\gamma}$ (normalized with the photon number) is generally assumed to be small (nanoscale) such that the net normalized lepton number equals the net baryon number $L = B$ where $B = [N_{\text{B}} - N_{\bar{\text{B}}}]/N_{\gamma}$. Barenboim, Kinney, and Park [121,122] note that the lepton asymmetry of the Universe is one of the most weakly constrained parameters is cosmology and they propose that models with leptogenesis are able to accommodate a large lepton number asymmetry surviving up to today.

If lepton number is grossly broken, this could provide a connection between cosmic neutrino properties and the baryon-antibaryon asymmetry present in the Universe today [122]. We quantify in [123] the impact of large lepton asymmetry on Universe expansion and show that there is another 'natural' choice $L \simeq 1$, making the net lepton number and net photon number in the Universe similar. Thus because N_{ν}^{eff} can be understood as a characterization of the relativistic dark radiation energy content in the early Universe, independent of its source, there still remains ambiguity in regard to measurements of N_{ν}^{eff}.

5. Electron-Positron Epoch

5.1. The Last Bastion of Antimatter

The electron-positron epoch of the early Universe was home to Big Bang Nucleosynthesis (BBN), the annihilation of most electrons and positrons reheating both the photon and neutrino fields, as well as setting the stage for the eventual recombination period which would generate the cosmic microwave background (CMB). The properties of the electron-positron $e^{\pm}$ plasma in the early Universe has not received appropriate attention in an era of precision BBN studies [124]. The presence of $e^{\pm}$ pairs before and during BBN has been acknowledged by Wang, Bertulani and Balantekin [125,126] over a decade ago. This however was before necessary tools were developed to explore the connection between electron and neutrino plasmas [25,38,109].

During the late stages of the $e^{\pm}$ epoch where BBN occurred, the matter content of the Universe was still mostly dominated by the light charged leptons by many orders of magnitude even though the Hubble parameter was still mostly governed by the radiation behavior of the neutrinos and photons. In Figure 19 we show that the dense $e^{\pm}$ plasma in the early Universe under the hypothesis charge neutrality and entropy conservation as a function of temperature $2\,\text{MeV} > T > 10\,\text{keV}$ [27]. The plasma is electron-positron rich, i.e., $n_{e^{\pm}} \gg n_B$ in the early Universe until leptonic annihilation at $T_{\text{split}} = 20.36$ keV. For $T < T_{\text{split}}$ the positron density n_{e^+} quickly vanishes because of annihilation leaving only a residual electron density as required by charge conservation.

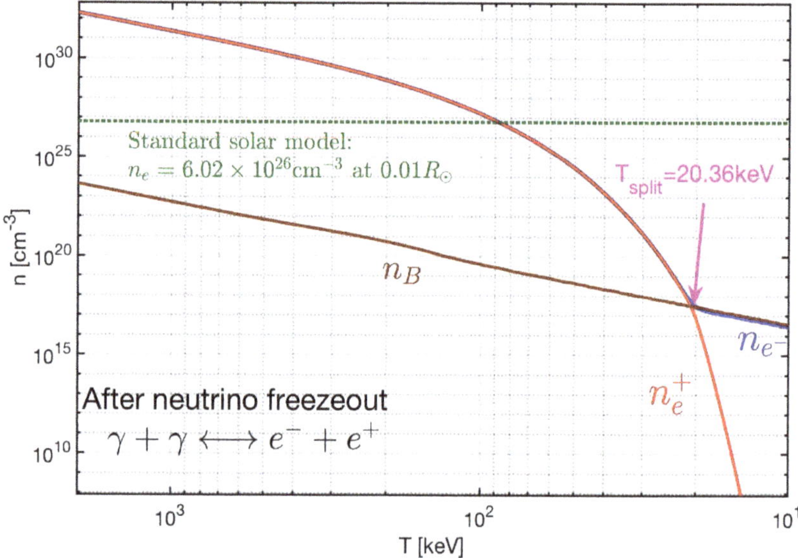

Figure 19. The $e^{\pm}$ number densities as a function of temperature in the range $2\,\text{MeV} > T > 10\,\text{keV}$. The blue solid line is the electron density n_{e^-}, the red solid line is the positron density n_{e^+}, and the brown solid line is the baryon density n_B. For comparison, we also show the green dotted line as the solar electron density within the solar core [127].

The temperatures during this epoch were also cool enough that the electrons and positrons could be described as partially non-relativistic to fairly good approximation while also still being as energy dense as the Solar core making it a relatively unique plasma environment not present elsewhere in cosmology. Considering the energy density between non-relativistic $e^{\pm}$ and baryons, we can write the ratio of energy densities as

$$\chi \equiv \frac{\rho_e + \rho_{\bar{e}}}{\rho_p + \rho_n}$$
$$= \frac{m_e(n_e + n_{\bar{e}})}{m_p n_p + m_n n_n} = \frac{m_e(n_e + n_{\bar{e}})}{n_B(m_p X_p + m_n X_n)} = \left(\frac{n_e + n_{\bar{e}}}{n_B}\right)\left(\frac{m_e}{m_p X_p + m_n X_\alpha/2}\right), \tag{74}$$

where we consider all neutrons as bound in ^4H_e after BBN. Species ratios $X_p = n_p/n_B$ and $X_\alpha = n_\alpha/n_B$ are given by the PDG [96] as

$$X_p = 0.878, \qquad X_\alpha = 0.245, \tag{75}$$

with masses

$$m_e = 0.511\,\text{MeV}, \qquad m_p = 938.272\,\text{MeV}, \qquad m_n = 939.565\,\text{MeV}. \tag{76}$$

In Figure 20 we plot the energy density ratio Equation (74) as a function of temperature $10 \, \text{keV} < T < 200 \, \text{keV}$. This figure shows that the energy density of electron and positron is dominant until $T = 28.2$ keV, i.e., at higher temperatures we have $\rho_e \gg \rho_B$. Until around $T \approx 85$ keV, the $e^\pm$ number density remained higher than that of the solar core, though notably at a much higher temperature than the Sun's core of $T_\odot = 1.36$ keV [51]. After $T = 28.2$ keV, where $\rho_e \ll \rho_B$, the ratio becomes constant around $T = 20$ keV because of positron annihilation and charge neutrality.

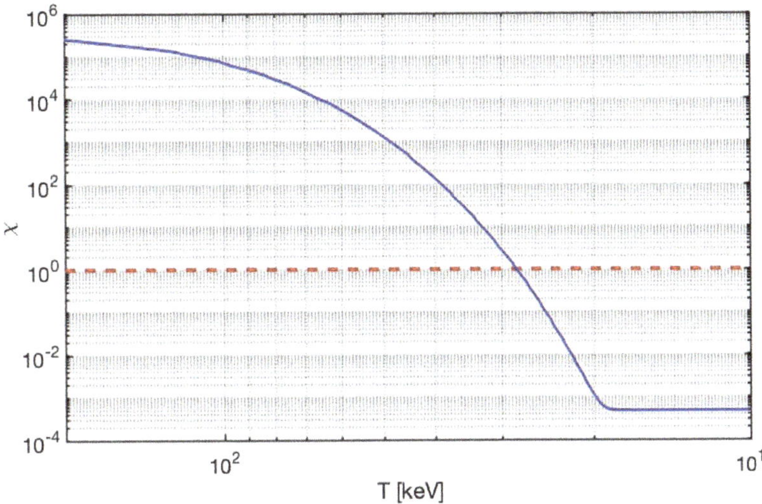

Figure 20. The energy density ratio χ (solid blue line) between $e^\pm$ and baryons as a function of temperature from $10 \, \text{keV} < T < 200 \, \text{keV}$. The dashed red line crossing point represents where the baryon density exceeds that of the electron-positron pairs.

5.2. Cosmic Magnetism

The Universe today filled with magnetic fields [128] at various scales and strengths both within galaxies and in deep extra-galactic space far and away from matter sources. Extra-galactic magnetic fields (EGMF) are not well constrained today, but are required by observation to be non-zero [129,130] with a magnitude between $10^{-12} \, \text{T} > B_{EGMF} > 10^{-20} \, \text{T}$ over Mpc coherent length scales. The upper bound is constrained from the characteristics of the CMB while the lower bound is constrained by non-observation of ultra-energetic photons from blazars [131]. There are generally considered two possible origins [132,133] for extra-galactic magnetic fields: (a) matter-induced dynamo processes involving Amperian currents and (b) primordial (or relic) seed magnetic fields whose origins may go as far back as the Big Bang itself. It is currently unknown which origin accounts for extra-galactic magnetic fields today or if it some combination of the two models. Even if magnetic fields in the Universe today are primarily driven via amplification through Amperian matter currents, such models could still benefit from the presence of primordial fields to act as catalyst. The purpose of this section is then to consider the magnetization properties of the $e^\pm$ plasma period due to spin which has not yet been considered.

While matter (and thus electrons) are relatively dilute today, the early Universe plasmas contained relatively large quantity of both matter (e^-) and antimatter (e^+). We explore here the spin response of the electron-positron plasma to external and self-magnetization fields thus developing methods for future detailed study.

As magnetic flux is conserved over co-moving surfaces, we see in Figure 21 that the primordial relic field is expected to dilute as $B \propto 1/a(t)^2$. This means the contemporary small bounded values of $5 \times 10^{-12} \, \text{T} > B_{relic} > 10^{-20} \, \text{T}$ (coherent over $\mathcal{O}$ (1 Mpc) distances) may have once represented large magnetic fields in the early Universe. Therefore, correctly describing the dynamics of this $e^{\pm}$ plasma is of interest when considering modern cosmic mysteries such as the origin of extra-galactic magnetic fields [129,131]. While most approaches tackle magnetized plasmas from the perspective of classical or semi-classical magneto-hydrodynamics (MHD) [134–137], our perspective is to demonstrate that fundamental quantum statistical analysis can lead to further insights on the behavior of magnetized plasmas.

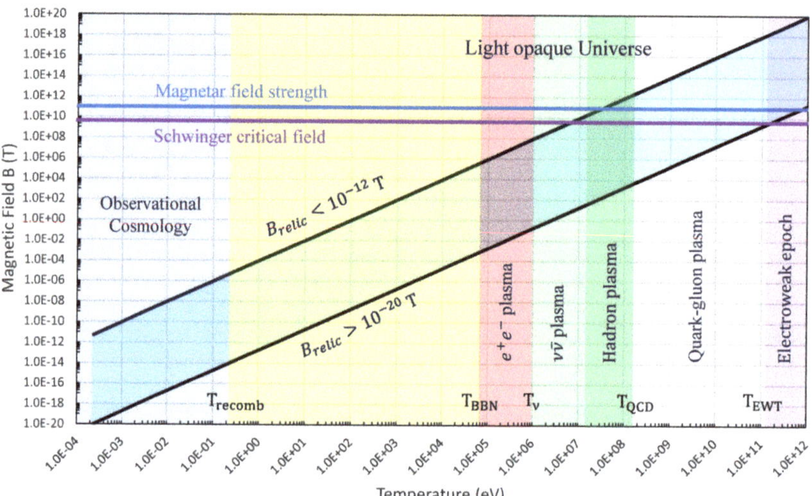

Figure 21. Qualitative value of the primordial magnetic field over the evolutionary lifespan of the Universe. The upper and lower black lines represent extrapolation of the EGMF bounds into the past. The major phases of the Universe are indicated with shaded regions. The values of the Schwinger critical field (purple line) and the upper bound of surface magnetar field strength (blue line) are included for scale.

As a starting point, we consider the energy eigenvalues of charged fermions within a homogeneous magnetic field. Here, we have several choices: We could assume the typical Dirac energy eigenvalues with gyro-magnetic g-factor set to $g = 2$. But as electrons, positrons and most plasma species have anomalous magnetic moments (AMM), we require a more complete model. Particle dynamics of classical particles with AMM are explored in [138–141]. Another option would be to modify the Dirac equation with a Pauli term [142], often called the Dirac-Pauli (DP) approach, via

$$\hat{H}_{\text{AMM}} = -a\frac{e}{2m_e}\frac{\sigma_{\mu\nu}F^{\mu\nu}}{2}, \tag{77}$$

where $\sigma_{\mu\nu}$ is the spin tensor proportional to the commutator of the gamma matrices and $F^{\mu\nu}$ is the EM field tensor. For the duration of this section, we will remain in natural units ($\hbar = c = k_B = 1$) unless explicitly stated otherwise. The AMM is defined via g-factor as

$$\frac{g}{2} = 1 + a. \tag{78}$$

This approach, while straightforward, would complicate the energies making analytic understanding and clarity difficult without a clear benefit. Modifying the Dirac equation with Equation (77) yields the following eigen-energies

$$
E_n^s\big|_{DP} = \sqrt{\left(\sqrt{m_e^2 + 2eB\left(n + \frac{1}{2} - s\right)} - \frac{eB}{2m}(g-2)s\right)^2 + p_z^2}
\tag{79}
$$

This model for the electron-positron plasma of the early Universe has been used in work such as Strickland et al. [143]. Our work in this section is then in part a companion piece which compares and contrasts the DP model of fermions to our preferred model for the AMM via the Klein-Gordon-Pauli (KGP) equation given by

$$
\left((i\partial_\mu - eA_\mu)^2 - m_e^2 - e\frac{g}{2}\frac{\sigma_{\mu\nu}F^{\mu\mu}}{2}\right)\Psi = 0.
\tag{80}
$$

We wish to emphasize, that each of the three above models (Dirac, DP, KGP) are distinct and have differing physical consequences and are not interchangeable which we explored in the context of hydrogen-like atoms in [144]. Recent work done in [145] discuss the benefits of KGP over other approaches for $g \neq 2$ from a quantum field theory perspective. Exploring the statistical behavior of KGP in a cosmological context can lead to new insights in magnetization which may be distinguished from pure $g = 2$ behavior of the Dirac equation or the *ad hoc* modification imposed by the Pauli term in DP. One major improvement of the KGP approach over the more standard DP approach is that the energies take eigenvalues which are mathematically similar to the Dirac energies. Considering the $e^\pm$ plasma in a uniform magnetic field B pointing along the z-axis, the energy of $e^\pm$ fermions can be written as

$$
\begin{aligned}
E_n^s &= \sqrt{p_z^2 + \tilde{m}^2 + 2eBn}, \\
\tilde{m}^2 &= m_e^2 + eB(1 - gs), \qquad s = \pm\tfrac{1}{2}, \qquad n = 0,1,2,3,\ldots
\end{aligned}
\tag{81}
$$

where n is the principle quantum number for the Landau levels and s is the spin quantum number. Here we introduce a notion of effective mass $\tilde{m}$ which inherits the spin-specific part of the energy adding them to the mass. This convention is also generalizable to further non-minimal electromagnetic models with more exotic energy contributions such that we write a general replacement as

$$
m_e^2 \to \tilde{m}^2(B).
\tag{82}
$$

This definition also pulls out the ground state Landau energy separating it from the remainder of the Landau tower of states. One restriction is that the effective mass must remain positive definite in our analysis thus we require

$$
\tilde{m}^2(B) = m_e^2 + eB(1 - gs) > 0.
\tag{83}
$$

This condition fails under ultra-strong magnetic fields of order

$$
B_{\text{crit}} = \frac{m_e^2}{ea} = \frac{\mathcal{B}_S}{a} \approx 3.8 \times 10^{12} \text{ T},
\tag{84}
$$

where $\mathcal{B}_S$ is the Schwinger critical field strength. For electrons, this field strength is well above the window of magnetic field strengths of interest during the late $e^\pm$ epoch.

5.3. Landau Eigen-Energies in Cosmology

There is another natural scale for the magnetic field besides Equation (84) when considering the consequences of FLRW expansion on the $e^\pm$ gas. As the Universe expands, different terms in the energies and thus partition function evolve as a function of the scale

of reduced entropy (maximum entropy yields the normal Fermi distribution itself with $\gamma = 1$) without pulling the system off a thermal temperature.

This situation is similar to that of the quarks during QGP, but instead the deviation is spatial rather than in time. This is precisely the kind of behavior that may arise in the $e^{\pm}$ epoch as the dominant photon thermal bath keeps the Fermi gas in thermal equilibrium while spatial inhomogeneity could spontaneously develop. For the remainder of this work, we will retain $\gamma(x) = 1$. The energy $E_n^{\pm}$ can be written as

$$E_n^{\pm} = \sqrt{p_z^2 + \tilde{m}_{\pm}^2 + 2eBn}, \qquad \tilde{m}_{\pm}^2 = m_e^2 + eB\left(1 \mp \frac{g}{2}\right),$$ (98)

where the $\pm$ script refers to spin aligned and anti-aligned eigenvalues. As we are interested in the temperature domain $T = 50$ keV, we can consider a semi-relativistic approach obtained by the Boltzmann approximation. Taking the limit $m_e/T \ll 1$, we obtain the first order Boltzmann approximation for semi-relativistic electrons and positrons. The Euler-Maclaurin formula is used to replace the sum over Landau levels with an integration which lets us split the partition function into three segments

$$\ln \mathcal{Z}_{tot} = \ln \mathcal{Z}_{free} + \ln \mathcal{Z}_B + \ln \mathcal{Z}_R,$$ (99)

where we define

$$\ln \mathcal{Z}_{free} = \frac{T^3 V}{2\pi^2} \sum_{i=\pm} \left[2\cosh\left(\frac{\eta_e^i}{T}\right)\right] x_i^2 K_2(x_i), \qquad x_i = \frac{\tilde{m}_i}{T}$$ (100)

$$\ln \mathcal{Z}_B = \frac{eBTV}{2\pi^2} \sum_{i=\pm} \left[2\cosh\left(\frac{\eta_e^i}{T}\right)\right]\left[\frac{x_i}{2}K_1(x_i) + \frac{b_0}{12}K_0(x_i)\right],$$ (101)

$$\ln \mathcal{Z}_R = \frac{eBTV}{\pi^2} \sum_{i=\pm} \left[2\cosh\left(\frac{\eta_e^i}{T}\right)\right] R.$$ (102)

The parameter R is the error remainder which is defined by integrals over Bernoulli polynomials. The parameter $\eta_e^{\pm}$ indicates that the chemical potential may be modified by the spin chemical potential and is in general non-zero as defined in Equation (97).

While this would require further derivation to demonstrate explicitly, the benefit of the Euler-Maclaurin approach is if the error contribution remains finite or bound for the magnetized partition function, then a correspondence between the free Fermi partition function (with noticeably modified effective mass $\tilde{m}_+$) and the magnetized Fermi partition function can be established. The mismatch between the summation and integral in the Euler-Maclaurin formula would then encapsulate the immediate magnetic response and deviation from the free particle phase space.

While we label $\ln \mathcal{Z}_{free}$ in Equation (100) as the "free" partition function, this is not strictly true as this contribution to the overall partition function is a function of the effective mass we defined earlier in Equation (82). When determining the magnetization of the quantum Fermi gas, derivatives of the magnetic field B will not fully vanish on this first term which will resulting in an intrinsic magnetization which is distinct from the contribution from the ground state and mismatch between the quantized Landau levels and the continuum of the free momentum. Specifically, this free Fermi contribution represents the magnetization that uniquely arises from the spin magnetic energy rather than orbital contributions.

Assuming the error remainder R is small and can be neglected, we can rewrite Equations (100) and (101) obtaining

$$\ln \mathcal{Z}_{tot} = \frac{T^3 V}{2\pi^2} \sum_{i=\pm} \left[2\cosh\left(\frac{\eta_e^i}{T}\right)\right]\left\{x_i^2 K_2(x_i) + \frac{b_0}{2}x_i K_1(x_i) + \frac{b_0^2}{12}K_0(x_i)\right\}.$$ (103)

Equation (103) is a surprisingly compact expression containing only tractable functions and will be our working model for the remainder of the work. Note that the above does not take into consideration density inhomogeneities and is restricted to the domain where the plasma is well described as a Maxwell-Boltzmann distribution. With that said, we have not taken the non-relativistic expansion of the eigen-energies.

5.5. Charge Neutrality and Chemical Potential

We explore the chemical potential of dense magnetized electron-positron plasma in the early Universe under the hypothesis of charge neutrality and entropy conservation. To learn about orders of magnitude we set in the following $\eta_e = \eta_e^+ = \eta_e^-$ and focus on the interval in the post-BBN temperature range $50\,\text{keV} > T > 20\,\text{keV}$. We return to the full problem under separate cover. The charge neutrality condition can be written as

$$(n_e - n_{\bar{e}}) = n_p = \left(\frac{n_p}{n_B}\right)\left(\frac{n_B}{s_{\gamma,\nu,e}}\right)s_{\gamma,\nu,e} = X_p\left(\frac{n_B}{s_{\gamma,\nu}}\right)s_{\gamma,\nu}, \qquad X_p \equiv \frac{n_p}{n_B}, \tag{104}$$

where n_p and n_B is the number density of protons and baryons respectively.

The radiation entropy component is given by $s_{\gamma,\nu}$. The entropy density contribution of $e^{\pm}$ is negligible compared to the photon and neutrino entropy density at post-BBN temperatures $50\,\text{keV} > T > 20\,\text{keV}$ because the low densities of $n_e \ll n_{\gamma,\nu}$ relative to the photon and neutrino gasses. The entropy density can be written as [69]

$$s = \frac{2\pi^2}{45}g_s T_\gamma^3, \qquad g_s = \sum_{i=boson} g_i\left(\frac{T_i}{T_\gamma}\right)^3 + \frac{7}{8}\sum_{i=fermion} g_i\left(\frac{T_i}{T_\gamma}\right)^3, \tag{105}$$

where g_s is the effective degree of freedom that contribute from boson and fermion species. The parameters X_p and (n_B/s) (see Equation (32)) can be determined by the observation, yielding $X_p = 0.878 \pm 0.015$ [96]. The net number density of electrons can be obtained by using the partition function of electron-positron plasma in the Boltzmann limit Equation (103) (with $g = 2$) as follows:

$$\begin{aligned}(n_e - n_{\bar{e}}) &= \frac{T}{V}\frac{\partial}{\partial \eta_e}\ln \mathcal{Z}_{tot} \\ &= \frac{T^3}{2\pi^2}[2\sinh(\eta_e/T)]\sum_{i=\pm}\left[x_i^2 K_2(x_i) + \frac{b_0}{2}x_i K_1(x_i) + \frac{b_0^2}{12}K_0(x_i)\right].\end{aligned} \tag{106}$$

Substituting Equation (106) into the charge neutrality condition Equation (104) we can solve the chemical potential of electron η_e/T yielding

$$\sinh(\eta_e/T) = \frac{2\pi^2}{2T^3}\frac{X_p(n_B/s_{\gamma,\nu})s_{\gamma,\nu}}{\sum_{i=\pm}\left[x_i^2 K_2(x_i) + \frac{b_0}{2}x_i K_1(x_i) + \frac{b_0^2}{12}K_0(x_i)\right]}, \tag{107}$$

$$\longrightarrow \frac{2\pi^2 n_p}{2T^3}\frac{X_p(n_B/s_{\gamma,\nu})s_{\gamma,\nu}}{2x^2 K_2(x)}, \qquad x = m_e/T, \qquad \text{for } b_0 = 0. \tag{108}$$

We see in Equation (108) that for the case $b_0 = 0$, the chemical potential agrees with the free particle result in [27].

5.6. Magnetization of the Electron-Positron Plasma

We consider the electron-positron plasma in the mean field approximation where the external field is representative of the "bulk" internal magnetization of the gas. Each particle is therefore responding to the averaged magnetic flux generated by its neighbors as well as any global external field contribution. Considering the magnetized electron-positron partition function Equation (103) we introduce dimensionless magnetization in S.I units and the critical field as follows

$$\frac{M}{H_c} = \frac{1}{H_c}\frac{k_B T}{V}\frac{\partial \ln \mathcal{Z}_{tot}}{\partial B}, \qquad H_c = \frac{B_c}{\mu_0} \qquad B_c = \frac{m_e^2 c^4}{e\hbar c^2}. \tag{109}$$

Applying Equation (109) to Equation (103) we arrive at the expression

$$M_\pm = \frac{eT^2}{2\pi^2}\left[2\cosh\left(\frac{\eta_e}{T}\right)\right]\{c_1(x_\pm)K_1(x_i) + c_0 K_0(x_\pm)\}, \tag{110}$$

$$c_1(x_\pm) = \left[\frac{1}{2} - \left(\frac{1}{2}\pm\frac{g}{4}\right)\left(1 + \frac{b_0^2}{12x_\pm^2}\right)\right]x_\pm, \qquad c_0 = \left[\frac{1}{6} - \left(\frac{1}{4}\pm\frac{g}{8}\right)\right]b_0. \tag{111}$$

Substituting the chemical potential Equation (107) into Equation (110) we can solve the magnetization M numerically. Considering the case $g = 2$ the magnetization can be written as the sum of the aligned and anti-aligned polarizations

$$M = M_+ + M_-, \tag{112}$$

where the functions $M_\pm$ are defined as

A. The aligned polarized gas is described by $\tilde{m}_+ = m_e$ and $x = \tilde{m}_+/T$. The magnetization of this contribution is therefore

$$M_+ = \frac{eT^2}{\pi^2}\sqrt{1+\sinh^2(\eta_e/T)}\left(\frac{1}{2}x_+ K_1(x_+) + \frac{b_0}{6}K_0(x_+)\right) \tag{113}$$

B. The spin anti-aligned gas has effective masses $\tilde{m}_- = \sqrt{m_e^2 + 2eB}$, and $x_- = \tilde{m}_-/T$. This yields a magnetization contribution of

$$M_- = -\frac{eT^2}{\pi^2}\sqrt{1+\sinh^2(\eta_e/T)}\left[\left(\frac{1}{2}+\frac{b_0^2}{12x_-^2}\right)x_- K_1(x_-) + \frac{b_0}{3}K_0(x_-)\right] \tag{114}$$

Using the cosmic magnetic scale parameter b_0 and chemical potential η_e/T we solve the magnetization numerically. In Figure 22, we present the outcome of this estimate. The solid lines (red for the lower bound of b_0 and blue for the higher bound of b_0) showing that the magnetization depends on the magnetic scale b_0.

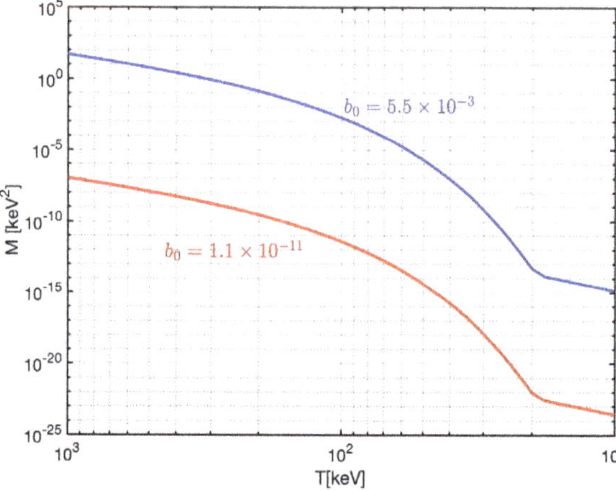

Figure 22. Estimate for the spin magnetization as a function of temperature in the range 10^3 keV $> T > 10$ keV, see text for detail.

6. Looking in the Cosmic Rear-View Mirror

The present day Universe seems devoid of antimatter but the primordial Universe was nearly matter-antimatter symmetric. There was only a fractional nano-scale excess of matter which today makes up the visible matter we see around us. All that remains of the tremendous initial amounts of matter-antimatter from the Big Bang is now seen as background thermal entropy. The origin of this nano-matter excess remains to this day an unresolved puzzle. If matter asymmetry emerged along the path of the Universe's evolution, as most think, the previously discussed Sakharov conditions (see Section 1.2) must be fulfilled.

We explored several major epochs in the Universe evolution where antimatter, in all its diverse forms, played a large roll. Emphasis was placed on understanding the thermal and chemical equilibria arising within the context of the Standard Model of particle physics. We highlighted that primordial quark-gluon plasma (QGP, which existed for $\approx 25\,\mu$s) is an important antimatter laboratory with its gargantuan antimatter content. Study of the QGP fireballs created in heavy-ion collisions performed today informs our understanding of the early Universe and vice versa [37,54,147,148], even though the primordial quark-gluon plasma under cosmic expansion explores a location in the phase diagram of QCD inaccessible to relativistic collider experiments considering both net baryon density, see Figure 5, and longevity of the plasma. We described (see Section 2.2) that the QGP epoch near to hadronization condition possessed bottom quarks in a non-equilibrium abundance: This novel QGP-Universe feature may be of interest in consideration of the QGP epoch as possible source for baryon asymmetry [63].

Bottom non-equilibrium is one among a few interesting results presented bridging the temperature gap between QGP hadronization at temperature $T \simeq 150\,$MeV and neutrino freeze-out. Specifically we shown **perseistnce of:**

- Strangeness abundance, present beyond the loss of the antibaryons at $T = 38.2\,$MeV.
- Pions, which are equilibrated via photon production long after the other hadrons disappear; these lightest hadrons are also dominating the Universe baryon abundance down to $T = 5.6\,$MeV.
- Muons, disappearing at around $T = 4.2\,$MeV, the condition when their decay rate outpaces their production rate.

At yet lower temperatures neutrinos make up the largest energy fraction in the Universe driving the radiation dominated cosmic expansion. Partway through this neutrino dominated Universe, in temperature range $T \in 3.5 - 1\,$MeV (range spanning differing flavor freeze-out, chemical equilibria, and even variation in standard natural constants; see Figure 18), the neutrinos freeze-out and decouple from the rest of the thermally active matter in the Universe. We consider neutrino decoupling condition as a function of elementary constants: If these constants were not all "constant" or significantly temperature dependent, a noticeable entropy flow of annihilating $e^{\pm}$ plasma into neutrinos could be present, generating additional so-called neutrino degrees of freedom.

We presented a detailed study of the evolving disappearance of the lightest antimatter, the positrons; we quantify the magnitude of the large positron abundance during and after Big Bang Nucleosynthesis (BBN), see Figure 19. In fact the energy density of electron-positron plasma exceeds greatly that of baryonic matter during and following the BBN period with the last positrons vanishing from the Universe near temperature $T = 20\,$keV, see Figure 20.

Looking forward, we note that some of the topics we explored deserve a more intense followup work:

- The study of matter baryogenesis in the context of bottom quarks chemical non-equilibrium persistence near to QGP hadronization;
- The impact of relatively dense $e^{\pm}$ plasma on BBN processes;
- Exploration of spatial inhomogeneities in dense $e^{\pm}$ plasma and eventual large scale structure formation and related spontaneous self magnetization process.

- Appearance of a significant positron abundance at $T > 25$ keV creates interest in understanding astrophysical object with core temperatures at, and beyond, this super-hot value; the high positron content enables in case of instability a rapid gamma ray formation akin to GRB events.

GRBs are current knowledge frontier: a tremendous amount of matter [6] must be converted into gammas in a short time-span of a few seconds. Ruffini and collaborators [4,7–9,11,149] suggests that strong field production of large amounts of antimatter which can be subsequently annihilated offers the most direct solution. This avoids the problem of excessive photon pressure needing to be balanced in super-hot objects where positron antimatter is already pre-existent. However, GRB events which lack classic after-signature supernova [150,151] could originate from novel super-hot stellar objects with primordial Universe properties which naturally possess, rather than create, larger amounts of positrons capable of rapid catalysis of gamma-rays upon gravitational collapse.

In conclusion: We hope that this work provides to all interested parties a first glimpse at the very interesting epoch of Universe evolution involving in sequence numerous plasma phases made of all particles known today. In this work we provided a background and connection for more specific periods found in the comprehensive literature of observational cosmology [152–156], the recombination period [28,157], BBN [124,158,159], and baryon asymmetry [96,160,161] or the origin of dark matter [30,162,163]. The Universe above temperatures $T > 130$ GeV and the inflation era [164,165] was outside the purview of this work.

Author Contributions: Conceived the article, supervised the development of research results, directed mode of presentation by J.R. equal weight by J.B.; A.S. and C.T.Y. Conceptualization: J.R.; software: J.B., A.S., C.T.Y.; validation: J.R., J.B., A.S., C.T.Y.; formal analysis: J.R., J.B., A.S., C.T.Y.; investigation: J.R., J.B., A.S., C.T.Y.; resources: J.R., J.B., A.S., C.T.Y.; data curation J.R., J.B., A.S., C.T.Y.; writing—original draft preparation: J.B., A.S., C.T.Y.; writing—review and editing: J.R.; A.S.; visualization: J.R., J.B., A.S., C.T.Y.; supervision: J.R.; project administration: J.R.; All authors have read and agreed to the published version of the manuscript.

Funding: This research received no external funding.

Data Availability Statement: This is a theoretical review, all results were converted into graphic display at creation and were in this format shared with the reader.

Acknowledgments:

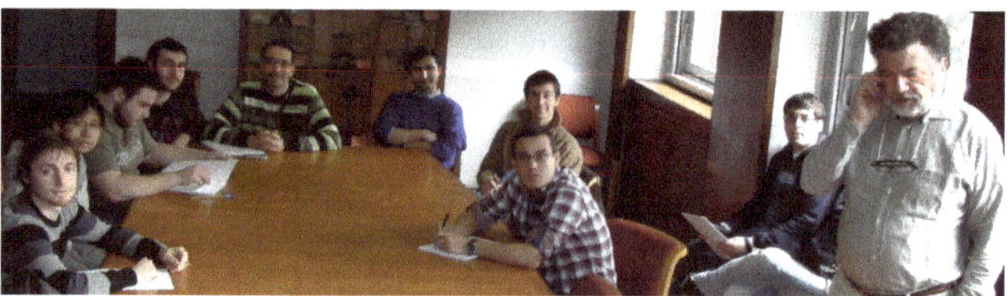

Figure 23. ICRANet group at work, Remo Ruffini on right. Photo by Johann Rafelski.

Figure 24. Lizhi Fang (on right), his wife Shuxian Li (center) and Shufang Su (Today: Physics Department Head at the University of Arizona) in April 2004. Photo taken by Johann Rafelski at his home in Tucson.

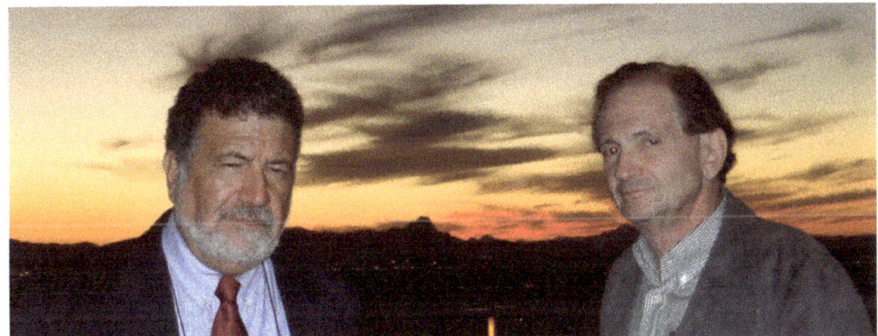

Figure 25. Remo Ruffini (on left) and Johann Rafelski beneath a sunset in Tucson, AZ on October 7th, 2012. The photo was taken by She Sheng Xue at a celebratory gathering honoring the life of Lizhi Fang.

This work was written in celebration of Professor Remo Ruffini's birthday, his contributions to astrophysics and cosmology and the large number of students and young scientists he mentored (see Figure 23). To close, we also acknowledge our mentor and colleague in the Department of Physics at the University of Arizona, Lizhi Fang [1–3] (see Figure 24) who passed away on April 6, 2012 at his home in Tucson, Arizona. Lizhi introduced Remo Ruffini to us, his career and life is remembered and celebrated (see Figure 25) as we continue to piece together the tapestry of the cosmos.

Conflicts of Interest: The authors declare no conflict of interest.

References

1. Fang, L.; Ruffini, R. (Eds.) Cosmology of the Early Universe. In *Advanced Series in Astrophysics and Cosmology*; World Scientific: Singapore, 1984; Volume 1.
2. Fang, L.; Ruffini, R. (Eds.) Galaxies, Quasars, and Cosmology. In *Advanced Series in Astrophysics and Cosmology*; World Scientific: Singapore, 1985; Volume 2.
3. Fang, L.; Ruffini, R. (Eds.) Quantum cosmology. In *Advanced Series in Astrophysics and Cosmology*; World Scientific: Singapore, 1987; Volume 3.
4. Ruffini, R.; Bianco, C.L.; Chardonnet, P.; Fraschetti, F.; Xue, S.S. On the interpretation of the burst structure of grbs. *Astrophys. J. Lett.* **2001**, *555*, L113–L116. [astro-ph/0106532]. [CrossRef]
5. Aksenov, A.G.; Bianco, C.L.; Ruffini, R.; Vereshchagin, G.V. GRBs and the thermalization process of electron-positron plasmas. *AIP Conf. Proc.* **2008**, *1000*, 309–312. [arXiv:astro-ph/0804.2807]. [CrossRef]

6. Aksenov, A.G.; Ruffini, R.; Vereshchagin, G.V. Pair plasma relaxation time scales. *Phys. Rev. E* **2010**, *81*, 046401. [arXiv:astro-ph.HE/1003.5616]. [CrossRef]

7. Ruffini, R.; Vereshchagin, G. Electron-positron plasma in GRBs and in cosmology. *Nuovo Cim. C* **2013**, *036*, 255–266. [arXiv:astro-ph.CO/1205.3512]. [CrossRef]

8. Ruffini, R.; Vitagliano, L.; Xue, S.S. Electron-positron-photon plasma around a collapsing star. In Proceedings of the 10th Marcel Grossmann Meeting on Recent Developments in Theoretical and Experimental General Relativity, Gravitation and Relativistic Field Theories (MG X MMIII), Rio de Janeiro, Brazil, 20–26 July 2003; pp. 295–302. [astro-ph/0304306]. [CrossRef]

9. Ruffini, R.; Vereshchagin, G.; Xue, S.S. Electron-positron pairs in physics and astrophysics: From heavy nuclei to black holes. *Phys. Rept.* **2010**, *487*, 1–140. [arXiv:astro-ph.HE/0910.0974]. [CrossRef]

10. Ruffini, R.; Salmonson, J.D.; Wilson, J.R.; Xue, S.S. On the pair-electromagnetic pulse from an electromagnetic black hole surrounded by a baryonic remnant. *Astron. Astrophys.* **2000**, *359*, 855. [astro-ph/0004257].

11. Han, W.B.; Ruffini, R.; Xue, S.S. Electron and positron pair production of compact stars. *Phys. Rev. D* **2012**, *86*, 84004. [arXiv:astro-ph.HE/1110.0700]. [CrossRef]

12. Belvedere, R.; Pugliese, D.; Rueda, J.A.; Ruffini, R.; Xue, S.S. Neutron star equilibrium configurations within a fully relativistic theory with strong, weak, electromagnetic, and gravitational interactions. *Nucl. Phys. A* **2012**, *883*, 1–24. [arXiv:astro-ph.SR/1202.6500]. [CrossRef]

13. Rafelski, J. Melting Hadrons, Boiling Quarks. *Eur. Phys. J. A* **2015**, *51*, 114. [arXiv:nucl-th/1508.03260]. [CrossRef]

14. Sakharov, A.D. Violation of CP Invariance, C asymmetry, and baryon asymmetry of the universe. *Pisma Zh. Eksp. Teor. Fiz.* **1967**, *5*, 32–35. [CrossRef]

15. Sakharov, A.D. Baryon asymmetry of the universe. *Sov. Phys. Uspekhi* **1991**, *34*, 417–421. [CrossRef]

16. Cohen, A.G.; De Rujula, A.; Glashow, S.L. A Matter-antimatter universe? *Astrophys. J.* **1998**, *495*, 539–549. [astro-ph/9707087]. [CrossRef]

17. Khlopov, M.Y.; Rubin, S.G.; Sakharov, A.S. Possible origin of antimatter regions in the baryon dominated universe. *Phys. Rev. D* **2000**, *62*, 083505. [hep-ph/0003285]. [CrossRef]

18. Blinnikov, S.I.; Dolgov, A.D.; Postnov, K.A. Antimatter and antistars in the universe and in the Galaxy. *Phys. Rev. D* **2015**, *92*, 023516. [arXiv:astro-ph.HE/1409.5736]. [CrossRef]

19. Khlopov, M.Y.; Lecian, O.M. The Formalism of Milky-Way Antimatter-Domains Evolution. *Galaxies* **2023**, *11*, 50. [CrossRef]

20. Affleck, I.; Dine, M. A New Mechanism for Baryogenesis. *Nucl. Phys. B* **1985**, *249*, 361–380. [CrossRef]

21. Rubakov, V.A.; Shaposhnikov, M.E. Electroweak baryon number nonconservation in the early universe and in high-energy collisions. *Usp. Fiz. Nauk* **1996**, *166*, 493–537. [hep-ph/9603208]. [CrossRef]

22. Letessier, J.; Rafelski, J. Hadron production and phase changes in relativistic heavy ion collisions. *Eur. Phys. J. A* **2008**, *35*, 221–242. [nucl-th/0504028]. [CrossRef]

23. Fromerth, M.J.; Kuznetsova, I.; Labun, L.; Letessier, J.; Rafelski, J. From Quark-Gluon Universe to Neutrino Decoupling: 200 < T < 2MeV. *Acta Phys. Polon. B* **2012**, *43*, 2261–2284. [arXiv:nucl-th/1211.4297]. [CrossRef]

24. Yang, C.T.; Rafelski, J. Cosmological strangeness abundance. *Phys. Lett. B* **2022**, *827*, 136944. [arXiv:hep-ph/2108.01752]. [CrossRef]

25. Birrell, J.; Yang, C.T.; Chen, P.; Rafelski, J. Relic neutrinos: Physically consistent treatment of effective number of neutrinos and neutrino mass. *Phys. Rev. D* **2014**, *89*, 23008. [arXiv:astro-ph.CO/1212.6943]. [CrossRef]

26. Birrell, J. Non-Equilibrium Aspects of Relic Neutrinos: From Freeze-out to the Present Day. Ph.D. Thesis, University of Arizona, Tucson, AZ, USA, 2014. [arXiv:nucl-th/1409.4500].

27. Grayson, C.; Yang, C.T.; Rafelski, J. Electron-Positron Plasma in the BBN epoch. 2023, *in preparation* .

28. Aghanim, N.; Akrami, Y.; Ashdown, M.; Aumont, J.; Baccigalupi, C.; Ballardini, M.; Banday, A.J.; Barreiro, R.B.; Bartolo, N.; Basak, S.; et al. Planck 2018 results. VI. Cosmological parameters. *Astron. Astrophys.* **2021**, *641*, A6. [arXiv:astro-ph.CO/1807.06209]. [CrossRef]

29. Rafelski, J.; Birrell, J. Traveling Through the Universe: Back in Time to the Quark-Gluon Plasma Era. *J. Phys. Conf. Ser.* **2014**, *509*, 012014. [arXiv:nucl-th/1311.0075]. [CrossRef]

30. Wantz, O.; Shellard, E.P.S. Axion Cosmology Revisited. *Phys. Rev. D* **2010**, *82*, 123508. [arXiv:astro-ph.CO/0910.1066]. [CrossRef]

31. Kronfeld, A.S. Lattice Gauge Theory and the Origin of Mass. In *100 Years of Subatomic Physics*; World Scientific: Singapore, 2013; pp. 493–518. [arXiv:physics.hist-ph/1209.3468]. [CrossRef]

32. D'Elia, M.; Mariti, M.; Negro, F. Susceptibility of the QCD vacuum to CP-odd electromagnetic background fields. *Phys. Rev. Lett.* **2013**, *110*, 082002. [arXiv:hep-lat/1209.0722]. [CrossRef]

33. Bonati, C.; Cossu, G.; D'Elia, M.; Mariti, M.; Negro, F. Effective θ term by CP-odd electromagnetic background fields. *PoS* **2014**, *LATTICE2013* , 360. [arXiv:hep-lat/1312.2805]. [CrossRef]

34. Borsanyi, S. Thermodynamics of the QCD transition from lattice. *Nucl. Phys. A* **2013**, *904–905*, 270c–277c. [arXiv:hep-lat/1210.6901]. [CrossRef]

35. Borsanyi, S.; Fodor, Z.; Hoelbling, C.; Katz, S.D.; Krieg, S.; Szabo, K.K. Full result for the QCD equation of state with 2 + 1 flavors. *Phys. Lett. B* **2014**, *730*, 99–104. [arXiv:hep-lat/1309.5258]. [CrossRef]

36. Bernstein, J. *Kinetic Theory in the Expanding Universe*; Cambridge Monographs on Mathematical Physics, Cambridge University Press: Cambridge, UK, 1988. [CrossRef]

37. Philipsen, O. The QCD equation of state from the lattice. *Prog. Part. Nucl. Phys.* **2013**, *70*, 55–107. [arXiv:hep-lat/1207.5999]. [CrossRef]
38. Mangano, G.; Miele, G.; Pastor, S.; Pinto, T.; Pisanti, O.; Serpico, P.D. Relic neutrino decoupling including flavor oscillations. *Nucl. Phys. B* **2005**, *729*, 221–234. [hep-ph/0506164]. [CrossRef]
39. Fornengo, N.; Kim, C.W.; Song, J. Finite temperature effects on the neutrino decoupling in the early universe. *Phys. Rev. D* **1997**, *56*, 5123–5134. [hep-ph/9702324]. [CrossRef]
40. Mangano, G.; Miele, G.; Pastor, S.; Peloso, M. A Precision calculation of the effective number of cosmological neutrinos. *Phys. Lett. B* **2002**, *534*, 8–16. [astro-ph/0111408]. [CrossRef]
41. Planck Collaboration. Planck 2013 results. XVI. Cosmological parameters. *Astron. Astrophys.* **2014**, *571*, A16. [arXiv:astro-ph.CO/1303.5076]. [CrossRef]
42. Planck Collaboration. Planck 2015 results. XIII. Cosmological parameters. *Astron. Astrophys.* **2016**, *594*, A13. [arXiv:astro-ph.CO/1502.01589]. [CrossRef]
43. Caldwell, R.R.; Kamionkowski, M.; Weinberg, N.N. Phantom energy and cosmic doomsday. *Phys. Rev. Lett.* **2003**, *91*, 071301. [astro-ph/0302506]. [CrossRef]
44. Bilic, N.; Tupper, G.B.; Viollier, R.D. Unification of dark matter and dark energy: The Inhomogeneous Chaplygin gas. *Phys. Lett. B* **2002**, *535*, 17–21. [astro-ph/0111325]. [CrossRef]
45. Benevento, G.; Hu, W.; Raveri, M. Can Late Dark Energy Transitions Raise the Hubble constant? *Phys. Rev. D* **2020**, *101*, 103517. [arXiv:astro-ph.CO/2002.11707]. [CrossRef]
46. Perivolaropoulos, L.; Skara, F. Challenges for ΛCDM: An update. *New Astron. Rev.* **2022**, *95*, 101659. [arXiv:astro-ph.CO/2105.05208]. [CrossRef]
47. Di Valentino, E.; Mena, O.; Pan, S.; Visinelli, L.; Yang, W.; Melchiorri, A.; Mota, D.F.; Riess, A.G.; Silk, J. In the realm of the Hubble tension—A review of solutions. *Class. Quant. Grav.* **2021**, *38*, 153001. [arXiv:astro-ph.CO/2103.01183]. [CrossRef]
48. Aluri, P.K.; Cea, P.; Chingangbam, P.; Chu, M.C.; Clowes, R.G.; Hutsemekers, D.; Kochappan, J.P.; Lopez, A.M.; Liu, L.; et al. Is the observable Universe consistent with the cosmological principle? *Class. Quant. Grav.* **2023**, *40*, 094001. [arXiv:astro-ph.CO/2207.05765]. [CrossRef]
49. Yan, H.; Ma, Z.; Ling, C.; Cheng, C.; Huang, J.S. First Batch of $z \approx 11$–20 Candidate Objects Revealed by the James Webb Space Telescope Early Release Observations on SMACS 0723-73. *Astrophys. J. Lett.* **2023**, *942*, L9. [arXiv:astro-ph.GA/2207.11558]. [CrossRef]
50. Weinberg, S. *Gravitation and Cosmology: Principles and Applications of the General Theory of Relativity*; John Wiley and Sons: New York, NY, USA, 1972.
51. Castellani, V.; Degl'Innocenti, S.; Fiorentini, G.; Lissia, M.; Ricci, B. Solar neutrinos: Beyond standard solar models. *Phys. Rept.* **1997**, *281*, 309–398. [astro-ph/9606180]. [CrossRef]
52. Rafelski, J.; Letessier, J.; Torrieri, G. Strange hadrons and their resonances: A Diagnostic tool of QGP freezeout dynamics. *Phys. Rev. C* **2002**, *64*, 054907. [CrossRef]
53. Ollitrault, J.Y. Anisotropy as a signature of transverse collective flow. *Phys. Rev. D* **1992**, *46*, 229–245. [CrossRef] [PubMed]
54. Petrán, M.; Letessier, J.; Petráček, V.; Rafelski, J. Hadron production and quark-gluon plasma hadronization in Pb Pb collisions at $\sqrt{s_{NN}} = 2.76$ TeV. *Phys. Rev. C* **2013**, *88*, 034907. [arXiv:hep-ph/1303.2098]. [CrossRef]
55. Ryu, S.; Paquet, J.F.; Shen, C.; Denicol, G.S.; Schenke, B.; Jeon, S.; Gale, C. Importance of the Bulk Viscosity of QCD in Ultrarelativistic Heavy-Ion Collisions. *Phys. Rev. Lett.* **2015**, *115*, 132301. [arXiv:nucl-th/1502.01675]. [CrossRef]
56. Rafelski, J. Formation and Observables of the Quark-Gluon Plasma. *Phys. Rept.* **1982**, *88*, 331.
57. Rafelski, J.; Schnabel, A. Quark-Gluon Plasma in Nuclear Collisions at 200-GeV/A. *Phys. Lett. B* **1988**, *207*, 6–10. [CrossRef]
58. Letessier, J.; Rafelski, J. *Hadrons and Quark–Gluon Plasma*; Cambridge Monographs on Particle Physics, Nuclear Physics and Cosmology, Cambridge University Press: Cambridge, UK, 2023. [CrossRef]
59. Bazavov, A.; Bhattacharya, T.; DeTar, C.; Ding, H.-T.; Gottlieb, S.; Gupta, R.; Hegde, P.; Heller, U.; Karsch, F.; Laermann, E.; et al. Equation of state in (2 + 1)-flavor QCD. *Phys. Rev. D* **2014**, *90*, 094503. [arXiv:hep-lat/1407.6387]. [CrossRef]
60. Jacak, B.V.; Muller, B. The exploration of hot nuclear matter. *Science* **2012**, *337*, 310–314. [CrossRef]
61. Fromerth, M.J.; Rafelski, J. *Hadronization of the quark Universe*; University of Arizona: Tucson, AZ, USA, 2005. [astro-ph/0211346].
62. Rafelski, J. Discovery of Quark-Gluon-Plasma: Strangeness Diaries. *Eur. Phys. J. ST* **2020**, *229*, 1–140. [arXiv:hep-ph/1911.00831]. [CrossRef]
63. Yang, C.T.; Rafelski, J. Possibility of bottom-catalyzed matter genesis near to primordial QGP hadronization. *arXiv* **2023**, arXiv:2004.06771. [arXiv:hep-ph/2004.06771].
64. Roberts, C.D. On Mass and Matter. *AAPPS Bull.* **2021**, *31*, 6. [arXiv:hep-ph/2101.08340]. [CrossRef]
65. Roberts, C.D. Origin of the Proton Mass. *EPJ Web Conf.* **2023**, *282*, 01006. [arXiv:hep-ph/2211.09905]. CrossRef]
66. Hagedorn, R. How We Got to QCD Matter from the Hadron Side: 1984. *Lect. Notes Phys.* **1985**, *221*, 53–76. [CrossRef]
67. Fromerth, M.J.; Rafelski, J. Limit on CPT violating quark anti-quark mass difference from the neutral kaon system. *Acta Phys. Polon. B* **2003**, *34*, 4151–4156. [hep-ph/0211362].
68. Tanabashi, M. et al. (Particle Data Group). Review of Particle Physics. *Phys. Rev. D* **2018**, *98*, 030001. [CrossRef]
69. Kolb, E.W.; Turner, M.S. *The Early Universe*; CRC Press: Boca Raton, FL, USA, 1990.

70. Kuznetsova, I.; Habs, D.; Rafelski, J. Pion and muon production in e−, e+, gamma plasma. *Phys. Rev. D* **2008**, *78*, 014027. [arXiv:hep-ph/0803.1588]. [CrossRef]

71. Kuznetsova, I.; Rafelski, J. Unstable Hadrons in Hot Hadron Gas in Laboratory and in the Early Universe. *Phys. Rev. C* **2010**, *82*, 035203. [arXiv:hep-th/1002.0375]. [CrossRef]

72. Rafelski, J.; Yang, C.T. The muon abundance in the primordial Universe. *Acta Phys. Polon. B* **2021**, *52*, 277. [arXiv:hep-ph/2103.07812]. [CrossRef]

73. Kuznetsova, I. Particle Production in Matter at Extreme Conditions. Master's Thesis, The University of Arizona, Tucson, AZ, USA, 2009. [arXiv:hep-th/0909.0524].

74. Kopp, J.; Maltoni, M.; Schwetz, T. Are There Sterile Neutrinos at the eV Scale? *Phys. Rev. Lett.* **2011**, *107*, 091801. [arXiv:hep-ph/1103.4570]. [CrossRef]

75. Hamann, J.; Hannestad, S.; Raffelt, G.G.; Wong, Y.Y.Y. Sterile neutrinos with eV masses in cosmology: How disfavoured exactly? *J. Cosmol. Astropart. Phys.* **2011**, *9*, 34. [arXiv:astro-ph.CO/1108.4136]. [CrossRef]

76. Kopp, J.; Machado, P.A.N.; Maltoni, M.; Schwetz, T. Sterile Neutrino Oscillations: The Global Picture. *J. High Energy Phys.* **2013**, *5*, 50. [arXiv:hep-ph/1303.3011]. [CrossRef]

77. Lello, L.; Boyanovsky, D. Cosmological Implications of Light Sterile Neutrinos produced after the QCD Phase Transition. *Phys. Rev. D* **2015**, *91*, 063502. [arXiv:astro-ph.CO/1411.2690]. [CrossRef]

78. Birrell, J.; Rafelski, J. Proposal for Resonant Detection of Relic Massive Neutrinos. *Eur. Phys. J. C* **2015**, *75*, 91. [arXiv:hep-ph/1402.3409]. [CrossRef]

79. Weinberg, S. Goldstone Bosons as Fractional Cosmic Neutrinos. *Phys. Rev. Lett.* **2013**, *110*, 241301. [arXiv:astro-ph.CO/1305.1971]. [CrossRef]

80. Giusarma, E.; Di Valentino, E.; Lattanzi, M.; Melchiorri, A.; Mena, O. Relic Neutrinos, thermal axions and cosmology in early 2014. *Phys. Rev. D* **2014**, *90*, 043507. [arXiv:astro-ph.CO/1403.4852]. [CrossRef]

81. Fukuda, Y.; Hayakawa, T.; Ichihara, E.; Inoue, K.; Ishihara, K.; Ishino, H.; Itow, Y.; Kajita, T.; Kameda, J.; Kasuga, S.; et al. Evidence for oscillation of atmospheric neutrinos. *Phys. Rev. Lett.* **1998**, *81*, 1562–1567. [hep-ex/9807003]. [CrossRef]

82. Eguchi, K.; Enomoto, S.; Furuno, K.; Goldman, J.; Hanada, H.; Ikeda, H.; Ikeda, K.; Inoue, K.; Ishihara, K.; Itoh, W.; et al. First results from KamLAND: Evidence for reactor antineutrino disappearance. *Phys. Rev. Lett.* **2003**, *90*, 21802. [hep-ex/0212021]. [CrossRef] [PubMed]

83. Fogli, G.L.; Lisi, E.; Marrone, A.; Palazzo, A. Global analysis of three-flavor neutrino masses and mixings. *Prog. Part. Nucl. Phys.* **2006**, *57*, 742–795. [hep-ph/0506083]. [CrossRef]

84. Giunti, C.; Studenikin, A. Neutrino electromagnetic interactions: A window to new physics. *Rev. Mod. Phys.* **2015**, *87*, 531. [arXiv:hep-ph/1403.6344]. [CrossRef]

85. Fritzsch, H.; Xing, Z.Z. Mass and flavor mixing schemes of quarks and leptons. *Prog. Part. Nucl. Phys.* **2000**, *45*, 1–81. [hep-ph/9912358]. [CrossRef]

86. Giunti, C.; Kim, C.W. *Fundamentals of Neutrino Physics and Astrophysics*; Oxford University Press: Oxford, UK, 2007.

87. Fritzsch, H. Neutrino Masses and Flavor Mixing. *Mod. Phys. Lett. A* **2015**, *30*, 1530012. [arXiv:hep-ph/1503.01857]. [CrossRef]

88. Dolinski, M.J.; Poon, A.W.P.; Rodejohann, W. Neutrinoless Double-Beta Decay: Status and Prospects. *Ann. Rev. Nucl. Part. Sci.* **2019**, *69*, 219–251. [arXiv:nucl-ex/1902.04097]. [CrossRef]

89. Arkani-Hamed, N.; Dimopoulos, S.; Dvali, G.R.; March-Russell, J. Neutrino masses from large extra dimensions. *Phys. Rev. D* **2001**, *65*, 024032. [hep-ph/9811448]. [CrossRef]

90. Ellis, J.R.; Lola, S. Can neutrinos be degenerate in mass? *Phys. Lett. B* **1999**, *458*, 310–321. [hep-ph/9904279]. [CrossRef]

91. Casas, J.A.; Ibarra, A. Oscillating neutrinos and $\mu \to e, \gamma$. *Nucl. Phys. B* **2001**, *618*, 171–204. [hep-ph/0103065]. [CrossRef]

92. King, S.F.; Luhn, C. Neutrino Mass and Mixing with Discrete Symmetry. *Rept. Prog. Phys.* **2013**, *76*, 56201. [arXiv:hep-ph/1301.1340]. [CrossRef]

93. Fernandez-Martinez, E.; Hernandez-Garcia, J.; Lopez-Pavon, J. Global constraints on heavy neutrino mixing. *J. High Energy Phys.* **2016**, *8*, 33. [arXiv:hep-ph/1605.08774]. [CrossRef]

94. Pascoli, S.; Petcov, S.T.; Riotto, A. Leptogenesis and Low Energy CP Violation in Neutrino Physics. *Nucl. Phys. B* **2007**, *774*, 1–52. [hep-ph/0611338]. [CrossRef]

95. Schwartz, M.D. *Quantum Field Theory and the Standard Model*; Cambridge University Press: Cambridge, UK, 2014.

96. Particle Data Group. Review of Particle Physics. *Prog. Theor. Exp. Phys.* **2022**, *2022*, 083C01. [CrossRef]

97. Avignone, F.T., III; Elliott, S.R.; Engel, J. Double Beta Decay, Majorana Neutrinos, and Neutrino Mass. *Rev. Mod. Phys.* **2008**, *80*, 481–516. [arXiv:nucl-ex/0708.1033]. [CrossRef]

98. Esteban, I.; Gonzalez-Garcia, M.C.; Maltoni, M.; Schwetz, T.; Zhou, A. The fate of hints: Updated global analysis of three-flavor neutrino oscillations. *J. High Energy Phys.* **2020**, *9*, 178. [arXiv:hep-ph/2007.14792]. [CrossRef]

99. Alvarez-Ruso, L.; Sajjad Athar, M.; Barbaro, M.B.; Cherdack, D.; Christy, M.E.; Coloma, P.; Donnelly, T.W.; Dytman, S.; de Gouvea, A.; Hill, R.J.; et al. NuSTEC White Paper: Status and challenges of neutrino–nucleus scattering. *Prog. Part. Nucl. Phys.* **2018**, *100*, 1–68. [arXiv:hep-ph/1706.03621]. [CrossRef]

100. Abi, B.; Acciarri, R.; Acero, M.A.; Adamov, G.; Adams, D.; Adinolfi, M.; Ahmad, Z.; Ahmed, J.; Alion, T.; Monsalve, S.A.; et al. Deep Underground Neutrino Experiment (DUNE), Far Detector Technical Design Report, Volume II: DUNE Physics. *arXiv* **2002**, arXiv:2002.03005. [arXiv:hep-ex/2002.03005].

101. Morgan, J.A. Cosmological upper limit to neutrino magnetic moments. *Phys. Lett. B* **1981**, *102*, 247–250. [CrossRef]
102. Fukugita, M.; Yazaki, S. Reexamination of Astrophysical and Cosmological Constraints on the Magnetic Moment of Neutrinos. *Phys. Rev. D* **1987**, *36*, 3817. [CrossRef] [PubMed]
103. Vogel, P.; Engel, J. Neutrino Electromagnetic Form-Factors. *Phys. Rev. D* **1989**, *39*, 3378. [CrossRef]
104. Elmfors, P.; Enqvist, K.; Raffelt, G.; Sigl, G. Neutrinos with magnetic moment: Depolarization rate in plasma. *Nucl. Phys. B* **1997**, *503*, 3–23. [hep-ph/9703214]. [CrossRef]
105. Giunti, C.; Studenikin, A. Neutrino electromagnetic properties. *Phys. Atom. Nucl.* **2009**, *72*, 2089–2125. [arXiv:hep-ph/0812.3646]. [CrossRef]
106. Canas, B.C.; Miranda, O.G.; Parada, A.; Tortola, M.; Valle, J.W.F. Updating neutrino magnetic moment constraints. *Phys. Lett. B* **2016**, *753*, 191–198. [arXiv:hep-ph/1510.01684]. [CrossRef]
107. Choquet-Bruhat, Y. *General Relativity and the Einstein Equations*; Oxford University Press: Oxford, UK, 2008.
108. Birrell, J.; Wilkening, J.; Rafelski, J. Boltzmann Equation Solver Adapted to Emergent Chemical Non-equilibrium. *J. Comput. Phys.* **2015**, *281*, 896–916. [arXiv:math.NA/1403.2019]. [CrossRef]
109. Birrell, J.; Yang, C.T.; Rafelski, J. Relic Neutrino Freeze-out: Dependence on Natural Constants. *Nucl. Phys. B* **2014**, *890*, 481–517. [arXiv:nucl-th/1406.1759]. [CrossRef]
110. Dreiner, H.K.; Hanussek, M.; Kim, J.S.; Sarkar, S. Gravitino cosmology with a very light neutralino. *Phys. Rev. D* **2012**, *85*, 65027. [arXiv:hep-ph/1111.5715]. [CrossRef]
111. Boehm, C.; Dolan, M.J.; McCabe, C. Increasing Neff with particles in thermal equilibrium with neutrinos. *J. Cosmol. Astropart. Phys.* **2012**, *12*, 27. [arXiv:astro-ph.CO/1207.0497]. [CrossRef]
112. Blennow, M.; Fernandez-Martinez, E.; Mena, O.; Redondo, J.; Serra, P. Asymmetric Dark Matter and Dark Radiation. *J. Cosmol. Astropart. Phys.* **2012**, *7*, 22. [arXiv:hep-ph/1203.5803]. [CrossRef]
113. Dicus, D.A.; Kolb, E.W.; Gleeson, A.M.; Sudarshan, E.C.G.; Teplitz, V.L.; Turner, M.S. Primordial Nucleosynthesis Including Radiative, Coulomb, and Finite Temperature Corrections to Weak Rates. *Phys. Rev. D* **1982**, *26*, 2694. [CrossRef]
114. Heckler, A.F. Astrophysical applications of quantum corrections to the equation of state of a plasma. *Phys. Rev. D* **1994**, *49*, 611–617. [CrossRef]
115. Mangano, G.; Miele, G.; Pastor, S.; Pinto, T.; Pisanti, O.; Serpico, P.D. Effects of non-standard neutrino-electron interactions on relic neutrino decoupling. *Nucl. Phys. B* **2006**, *756*, 100–116. [hep-ph/0607267]. [CrossRef]
116. Anchordoqui, L.A.; Goldberg, H. Neutrino cosmology after WMAP 7-Year data and LHC first Z' bounds. *Phys. Rev. Lett.* **2012**, *108*, 81805. [arXiv:hep-ph/1111.7264]. [CrossRef]
117. Abazajian, K.N.; Acero, M.A.; Agarwalla, S.K.; Aguilar-Arevalo, A.A.; Albright, C.H.; Antusch, S.; Arguelles, C.A.; Balantekin, A.B.; Barenboim, G.; Barger, V.; et al. Light Sterile Neutrinos: A White Paper. *arXiv* **2012**, arXiv:1204.5379. [arXiv:hep-ph/1204.5379].
118. Anchordoqui, L.A.; Goldberg, H.; Steigman, G. Right-Handed Neutrinos as the Dark Radiation: Status and Forecasts for the LHC. *Phys. Lett. B* **2013**, *718*, 1162–1165. [arXiv:hep-ph/1211.0186]. [CrossRef]
119. Steigman, G. Equivalent Neutrinos, Light WIMPs, and the Chimera of Dark Radiation. *Phys. Rev. D* **2013**, *87*, 103517. [arXiv:astro-ph.CO/1303.0049]. [CrossRef]
120. Birrell, J.; Rafelski, J. Quark–gluon plasma as the possible source of cosmological dark radiation. *Phys. Lett. B* **2015**, *741*, 77–81. [arXiv:nucl-th/1404.6005]. [CrossRef]
121. Barenboim, G.; Kinney, W.H.; Park, W.I. Resurrection of large lepton number asymmetries from neutrino flavor oscillations. *Phys. Rev. D* **2017**, *95*, 43506. [arXiv:hep-ph/1609.01584]. [CrossRef]
122. Barenboim, G.; Park, W.I. A full picture of large lepton number asymmetries of the Universe. *J. Cosmol. Astropart. Phys.* **2017**, *4*, 48. [arXiv:hep-ph/1703.08258]. [CrossRef]
123. Yang, C.T.; Birrell, J.; Rafelski, J. Lepton Number and Expansion of the Universe. *arXiv* **2018**, arXiv:1812.05157. [arXiv:hep-ph/1812.05157].
124. Pitrou, C.; Coc, A.; Uzan, J.P.; Vangioni, E. Precision big bang nucleosynthesis with improved Helium-4 predictions. *Phys. Rept.* **2018**, *754*, 1–66. [arXiv:astro-ph.CO/1801.08023]. [CrossRef]
125. Wang, B.; Bertulani, C.A.; Balantekin, A.B. Electron screening and its effects on Big-Bang nucleosynthesis. *Phys. Rev. C* **2011**, *83*, 18801. [arXiv:astro-ph.CO/1010.1565]. [CrossRef]
126. Hwang, E.; Jang, D.; Park, K.; Kusakabe, M.; Kajino, T.; Balantekin, A.B.; Maruyama, T.; Ryu, C.M.; Cheoun, M.K. Dynamical screening effects on big bang nucleosynthesis. *J. Cosmol. Astropart. Phys.* **2021**, *11*, 17. [arXiv:nucl-th/2102.09801]. [CrossRef]
127. Bahcall, J.N.; Pinsonneault, M.H.; Basu, S. Solar models: Current epoch and time dependences, neutrinos, and helioseismological properties. *Astrophys. J.* **2001**, *555*, 990–1012. [astro-ph/0010346]. [CrossRef]
128. Kronberg, P.P. Extragalactic magnetic fields. *Rept. Prog. Phys.* **1994**, *57*, 325–382. [CrossRef]
129. Anchordoqui, L.A.; Goldberg, H. A Lower bound on the local extragalactic magnetic field. *Phys. Rev. D* **2002**, *65*, 21302. [hep-ph/0106217]. [CrossRef]
130. Widrow, L.M. Origin of galactic and extragalactic magnetic fields. *Rev. Mod. Phys.* **2002**, *74*, 775–823. [astro-ph/0207240]. [CrossRef]
131. Neronov, A.; Vovk, I. Evidence for strong extragalactic magnetic fields from Fermi observations of TeV blazars. *Science* **2010**, *328*, 73–75. [arXiv:astro-ph.HE/1006.3504]. [CrossRef]

132. Widrow, L.M.; Ryu, D.; Schleicher, D.R.G.; Subramanian, K.; Tsagas, C.G.; Treumann, R.A. The First Magnetic Fields. *Space Sci. Rev.* **2012**, *166*, 37–70. [arXiv:astro-ph.CO/1109.4052]. [CrossRef]

133. Vazza, F.; Locatelli, N.; Rajpurohit, K.; Banfi, S.; Dominguez-Fernandez, P.; Wittor, D.; Angelinelli, M.; Inchingolo, G.; Brienza, M.; Hackstein, S.; et al. Magnetogenesis and the Cosmic Web: A Joint Challenge for Radio Observations and Numerical Simulations. *Galaxies* **2021**, *9*, 109. [arXiv:astro-ph.CO/2111.09129]. [CrossRef]

134. Berezhiani, V.; Tsintsadze, L.; Shukla, P. Influence of electron-positron pairs on the wakefields in plasmas. *Phys. Scr.* **1992**, *46*, 55. [CrossRef]

135. Berezhiani, V.; Mahajan, S. Large relativistic density pulses in electron-positron-ion plasmas. *Phys. Rev. E* **1995**, *52*, 1968. [CrossRef]

136. Schlickeiser, R.; Kolberg, U.; Yoon, P.H. Primordial Plasma Fuctuations. I. Magnetization of the Early Universe by Dark Aperiodic Fluctuations in the Past Myon and Prior Electron–Positron Annihilation Epoch. *Astrophys. J.* **2018**, *857*, 29. [CrossRef]

137. Melrose, D. *Quantum Plasmadynamics: Magnetized Plasmas*; Springer: Berlin/Heidelberg, Germany, 2013. [CrossRef]

138. Rafelski, J.; Formanek, M.; Steinmetz, A. Relativistic Dynamics of Point Magnetic Moment. *Eur. Phys. J. C* **2018**, *78*, 6. [arXiv:physics.class-ph/1712.01825]. [CrossRef]

139. Formanek, M.; Evans, S.; Rafelski, J.; Steinmetz, A.; Yang, C.T. Strong fields and neutral particle magnetic moment dynamics. *Plasma Phys. Control. Fusion* **2018**, *60*, 74006. [arXiv:hep-ph/1712.07698]. [CrossRef]

140. Formanek, M.; Steinmetz, A.; Rafelski, J. Radiation reaction friction: Resistive material medium. *Phys. Rev. D* **2020**, *102*, 56015. [arXiv:hep-ph/2004.09634]. [CrossRef]

141. Formanek, M.; Steinmetz, A.; Rafelski, J. Motion of classical charged particles with magnetic moment in external plane-wave electromagnetic fields. *Phys. Rev. A* **2021**, *103*, 52218. [arXiv:physics.class-ph/2103.02594]. [CrossRef]

142. Thaller, B. *The Dirac Equation*; Springer Science & Business Media: Berlin/Heidelberg, Germany, 2013.

143. Strickland, M.; Dexheimer, V.; Menezes, D.P. Bulk Properties of a Fermi Gas in a Magnetic Field. *Phys. Rev. D* **2012**, *86*, 125032. [arXiv:nucl-th/1209.3276]. [CrossRef]

144. Steinmetz, A.; Formanek, M.; Rafelski, J. Magnetic Dipole Moment in Relativistic Quantum Mechanics. *Eur. Phys. J. A* **2019**, *55*, 40. [arXiv:hep-ph/1811.06233]. [CrossRef]

145. Rafelski, J.; Evans, S.; Labun, L. Study of QED singular properties for variable gyromagnetic ratio $g \simeq 2$. *Phys. Rev. D* **2023**, *107*, 76002. [arXiv:hep-th/2212.13165]. [CrossRef]

146. Elze, H.T.; Greiner, W.; Rafelski, J. The relativistic ideal Fermi gas revisited. *J. Phys. G* **1980**, *6*, L149–L153. [CrossRef]

147. Borsanyi, S.; Fodor, Z.; Guenther, J.; Kampert, K.-H.; Katz, S.D.; Kawanai, T.; Kovacs, T.G.; Mages, S.W.; Pasztor, A.; Pittler, F.; et al. Calculation of the axion mass based on high-temperature lattice quantum chromodynamics. *Nature* **2016**, *539*, 69–71. [arXiv:hep-lat/1606.07494]. [CrossRef]

148. Rafelski, J. Connecting QGP-Heavy Ion Physics to the Early Universe. *Nucl. Phys. B Proc. Suppl.* **2013**, *243–244*, 155–162. [arXiv:astro-ph.CO/1306.2471]. [CrossRef]

149. Aksenov, A.G.; Ruffini, R.; Vereshchagin, G.V. Thermalization of electron-positron-photon plasmas with an application to GRB. *AIP Conf. Proc.* **2008**, *966*, 191–196. [CrossRef]

150. Burns, E.; Svinkin, D.; Fenimore, E.; Kann, D.A.; Fernandez, J.F.A.; Frederiks, D.; Hamburg, R.; Lesage, S.; Temiraev, Y.; Tsvetkova, A.; et al. GRB 221009A: The BOAT. *Astrophys. J. Lett.* **2023**, *946*, L31. [arXiv:astro-ph.HE/2302.14037]. [CrossRef]

151. Levan, A.J.; Lamb, G.P.; Schneider, B.; Hjorth, J.; Zafar, T.; Postigo, A.D.; Sargent, B.; Mullally, S.E.; Izzo, L.; D'Avanzo, P.; et al. The First JWST Spectrum of a GRB Afterglow: No Bright Supernova in Observations of the Brightest GRB of all Time, GRB 221009A. *Astrophys. J. Lett.* **2023**, *946*, L28. [arXiv:astro-ph.HE/2302.07761]. [CrossRef]

152. Davis, M.; Efstathiou, G.; Frenk, C.S.; White, S.D.M. The Evolution of Large Scale Structure in a Universe Dominated by Cold Dark Matter. *Astrophys. J.* **1985**, *292*, 371–394. [CrossRef]

153. Navarro, J.F.; Frenk, C.S.; White, S.D.M. The Structure of cold dark matter halos. *Astrophys. J.* **1996**, *462*, 563–575. [astro-ph/9508025]. [CrossRef]

154. Moore, B.; Ghigna, S.; Governato, F.; Lake, G.; Quinn, T.R.; Stadel, J.; Tozzi, P. Dark matter substructure within galactic halos. *Astrophys. J. Lett.* **1999**, *524*, L19–L22. [astro-ph/9907411]. [CrossRef]

155. Springel, V.; White, S.D.M.; Jenkins, A.; Frenk, C.S.; Yoshida, N.; Gao, L.; Navarro, J.; Thacker, R.; Croton, D.; Helly, J.; et al. Simulating the joint evolution of quasars, galaxies and their large-scale distribution. *Nature* **2005**, *435*, 629–636. [astro-ph/0504097]. [CrossRef] [PubMed]

156. Arbey, A.; Mahmoudi, F. Dark matter and the early Universe: A review. *Prog. Part. Nucl. Phys.* **2021**, *119*, 103865. [arXiv:hep-ph/2104.11488]. [CrossRef]

157. Planck Collaboration. Planck 2018 results. I. Overview and the cosmological legacy of Planck. *Astron. Astrophys.* **2020**, *641*, A1. [arXiv:astro-ph.CO/1807.06205]. [CrossRef]

158. Steigman, G. Primordial Nucleosynthesis in the Precision Cosmology Era. *Ann. Rev. Nucl. Part. Sci.* **2007**, *57*, 463–491. [arXiv:astro-ph/0712.1100]. [CrossRef]

159. Cyburt, R.H.; Fields, B.D.; Olive, K.A.; Yeh, T.H. Big Bang Nucleosynthesis: 2015. *Rev. Mod. Phys.* **2016**, *88*, 015004. [arXiv:astro-ph.CO/1505.01076]. [CrossRef]

160. Kuzmin, V.A.; Rubakov, V.A.; Shaposhnikov, M.E. On the Anomalous Electroweak Baryon Number Nonconservation in the Early Universe. *Phys. Lett. B* **1985**, *155*, 36. [CrossRef]

161. Canetti, L.; Drewes, M.; Shaposhnikov, M. Matter and Antimatter in the Universe. *New J. Phys.* **2012**, *14*, 95012. [arXiv:hep-ph/1204.4186]. [CrossRef]
162. Bertone, G.; Hooper, D.; Silk, J. Particle dark matter: Evidence, candidates and constraints. *Phys. Rept.* **2005**, *405*, 279–390. [hep-ph/0404175]. [CrossRef]
163. Peccei, R.D. The Strong CP problem and axions. *Lect. Notes Phys.* **2008**, *741*, 3–17. [hep-ph/0607268]. [CrossRef]
164. Baumann, D. Inflation. In Proceedings of the Theoretical Advanced Study Institute in Elementary Particle Physics: Physics of the Large and the Small, Boulder, CO, USA, 1–26 June 2009; pp. 523–686. [arXiv:hep-th/0907.5424]. [CrossRef]
165. Allahverdi, R.; Amin, M.A.; Berlin, A.; Bernal, N.; Byrnes, C.T.; Delos, M.S.; Erickcek, A.L.; Escudero, M.; Figueroa, D.G.; Freese, K.; et al. The First Three Seconds: A Review of Possible Expansion Histories of the Early Universe. *Open J. Astrophys.* **2021**, *4*, 1. [arXiv:astro-ph.CO/2006.16182]. [CrossRef]

Article

A Wheeler–DeWitt Quantum Approach to the Branch-Cut Gravitation with Ordering Parameters

Benno August Ludwig Bodmann [1], César Augusto Zen Vasconcellos [2,3,*], Peter Otto Hess Bechstedt [4,5], José Antonio de Freitas Pacheco [6], Dimiter Hadjimichef [2], Moisés Razeira [7] and Gervásio Annes Degrazia [1]

[1] Departamento de Física, Universidade Federal de Santa Maria (UFSM), Santa Maria 97105-900, Brazil; benno.bodmann@gmx.de (B.A.L.B.); gervasiodegrazia@gmail.com (G.A.D.)
[2] Instituto de Física, Universidade Federal do Rio Grande do Sul (UFRGS), Porto Alegre 90010-150, Brazil; dimiter@ufrgs.br
[3] International Center for Relativistic Astrophysics Network (ICRANet), 65122 Pescara, Italy
[4] Instituto de Ciencias Nucleares, Universidad Nacional Autónoma de Mexico (UNAM), Mexico City 04510, Mexico; hess@nucleares.unam.mx
[5] Frankfurt Institute for Advanced Studies (FIAS), 60438 Hessen, Germany
[6] Observatoire de la Côte d'Azur, 06300 Nice, France; jose.pacheco@oca.eu
[7] Laboratório de Geociências Espaciais e Astrofísica (LaGEA), Universidade Federal do Pampa (UNIPAMPA), Caçapava do Sul 96570-000, Brazil; moisesrazeira@unipampa.edu.br
* Correspondence: cesarzen@cesarzen.com; Tel.: +55-(51)-98357-1902

Citation: Bodmann, B.A.L.; Vasconcellos, C.A.Z.; Bechstedt, P.O.H.; de Freitas Pacheco, J.A.; Hadjimichef, D.; Razeira, M.; Degrazia, G.A. A Wheeler–DeWitt Quantum Approach to the Branch-Cut Gravitation with Ordering Parameters. *Universe* **2023**, *9*, 278. https://doi.org/10.3390/universe9060278

Academic Editors: Gregory Vereshchagin, Narek Sahakyan and Jorge Armando Rueda Hernández

Received: 4 May 2023
Revised: 30 May 2023
Accepted: 6 June 2023
Published: 8 June 2023

Abstract: In this contribution to the Festschrift for Prof. Remo Ruffini, we investigate a formulation of quantum gravity using the Hořava–Lifshitz theory of gravity, which is General Relativity augmented by counter-terms to render the theory regularized. We are then led to the Wheeler–DeWitt (WDW) equation combined with the classical concepts of the branch-cut gravitation, which contemplates as a new scenario for the origin of the Universe, a smooth transition region between the contraction and expansion phases. Through the introduction of an energy-dependent effective potential, which describes the space-time curvature associated with the embedding geometry and its coupling with the cosmological constant and matter fields, solutions of the WDW equation for the wave function of the Universe are obtained. The Lagrangian density is quantized through the standard procedure of raising the Hamiltonian, the helix-like complex scale factor of branched gravitation as well as the corresponding conjugate momentum to the category of quantum operators. Ambiguities in the ordering of the quantum operators are overcome with the introduction of a set of ordering factors α, whose values are restricted, to make contact with similar approaches, to the integers $\alpha = [0, 1, 2]$, allowing this way a broader class of solutions for the wave function of the Universe. In addition to a branched universe filled with underlying background vacuum energy, primordial matter and radiation, in order to connect with standard model calculations, we additionally supplement this formulation with baryon matter, dark matter and quintessence contributions. Finally, the boundary conditions for the wave function of the Universe are imposed by assuming the Bekenstein criterion. Our results indicate the consistency of a topological quantum leap, or alternatively a quantum tunneling, for the transition region of the early Universe in contrast to the classic branched cosmology view of a smooth transition.

Keywords: branch-cut cosmology; Wheeler–DeWitt equation; quantum gravity

1. Introduction

Motivated by the success of quantum mechanics (QM) and pseudo-complex general relativity (pc-GR) in incorporating the mathematics principles of existential closure and completeness [1] by extending their domains of realization, QM to the complex variables sector [2], and pc-GR to the pseudo-complex domain [3–5], branch-cut gravitation theory (BCGT), in its classical version, represents an analytically continued extension of general relativity [6] to the complex plane [7–13]. The descriptive augmented domain of quantum mechanics by incorporating complex variables has broadened our perception of the infinitesimally small scales, with direct physical manifestations [14,15]. In turn, these notions led pc-GR, embedded in a pseudo-complex domain, to a suppression mechanism of the primordial gravitational singularity and to the prediction of existence of dark energy outside and inside cosmic mass distributions [3–5].

BCGT describes a hypothetical set of independent multiple universes existing in parallel, based on the multiverse[1] conception by Hawking and Hertog [16], each emerging from its own singularity. Imposing that the multiverses compose a single universe, in the Riemann limit, the multiple singularities merge, generating topological and complex smooth structures of foliation leafs, continuously connected, described by Riemann surfaces. The corresponding solutions of the analitically continued Einstein equations, represented by the helix-shaped branch-cut function $\ln^{-1}[\beta(t)]$, give rise to an alternative formulation of the Friedmann equations, as a function of complex time t, given by[2]

$$\left(\frac{\frac{d}{dt}\ln^{-1}[\beta(t)]}{\ln^{-1}[\beta(t)]} \right)^2 = \frac{8\pi G}{3}\rho(t) - \frac{kc^2}{\ln^{-1}[\beta(t)]}, \tag{1}$$

and

$$\left(\frac{\frac{d^2}{dt^2}\ln^{-1}[\beta(t)]}{\ln^{-1}[\beta(t)]} \right) = -\frac{4\pi G}{3}\left(\rho(t) + \frac{3}{c^2}p(t) \right). \tag{2}$$

Equations (1) and (2), and their corresponding complex conjugate versions, describe a smooth universe with a fine-tune transition region from contraction to expansion — purely geometric in nature, that replaces the cosmological singularity (Figure 1). Similar procedures allow to obtain analytically continued expressions for the energy-stress conservation law, Hubble rate, deceleration parameter, Ricci scalar and the Ricci curvature, as well as the corresponding complex conjugated expressions.

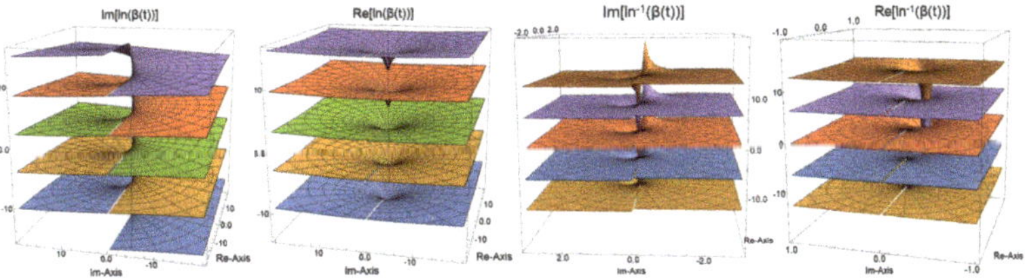

Figure 1. On the two left figures, characteristic plots of the Riemann surface associated with the imaginary and the real parts of the function $\ln[\beta(t)]$, the scaling in time of the branch-cut universe (the reciprocal of $\ln^{-1}[\beta(t)]$). The plot of the imaginary part shows connected glued domains: the various branches of the function are glued along the copies of each upper half plane with their copies on the corresponding lower half plane in a suitable way to make $\ln[\beta(t)]$ continuous. Each time the variable β moves around the origin, $\ln[\beta(t)]$ moves to a different branch, with its values, on each foliaition leaf, differing from its principal value by a multiple of $2\pi i$. A similar analysis apply to $\ln^{-1}[\beta(t)]$. On the two right figures, characteristic plots of $\ln^{-1}[\beta(t)]$.

In branching gravitation, the primordial singularity is replaced by a family of Riemann foliation leafs in which the branch-cut cosmic scale factor[3] $\ln^{-1}[\beta(t)]$ shrinks to a finite critical size, shaped by the range-, foliation leafs regularization- and domain extension-$\beta(t)$-function, with its range domain above the Planck length according to the Bekenstein criterion[4] [11]. In the contraction phase, as the patch size decreases with a linear dependence on $\ln[\beta(t)]$, light travels through geodesics on each Riemann foliation leaf, circumventing continuously the branch-cut, and although the horizon size scale with $\ln^{\epsilon}[\beta(t)]/\ln[\beta(t)]$, where ϵ denotes the dimensionless thermodynamics connection, the length of the path to be traveled by light compensates for the scaling difference between the patch and horizon sizes. Under these conditions, causality between the horizon size and the patch size may be achieved through the accumulation of branches in the transition region between the present state of the universe and the past events [11]. In addition to causality, the flatness and the horizon dilemmas of cosmology stand out. The flatness problem concerns the value of the ratio between the total density of the universe and the critical density resulting in a very small Planck value of the time-dependent and dimensionless cosmic spatial factor [17–19], Ω_c, which scales as $\ln^{2\epsilon}[\beta(t)]/\ln^2[\beta(t)]$. The horizon problem in turn arises exactly because the patch corresponding to the observable universe was never causally connected in the past [17–19]. The restoration of causality in BCGT brings an additional perspective with a view to the future elucidation of these 'cosmological puzzles' [12].

In short, in the BCGT formalism of spacetime, a 3 + 1 dimensional Riemannian manifold $\mathcal{M}$ with metric g, is foliated into a one parameter family of space-like slices (leafs) or continuous trajetories (see Figure 1), with the spatial slices assumed to be closed. As a corolary, the branched gravitation approach only expands the domain of realization of the governing principles of general relativity, as well as the operations that underlie its theoretical foundations.

Recently, we have proposed a topological canonical quantum approach [13] for the classical branch-cut cosmology on basis of the renormalizable Hořava–Lifshitz theory of gravity (HLGT) [20] and the Wheeler–DeWitt Equation (WdW) [21]. HLGT is General Relativity augmented by counter-terms to render the theory regularized[5]. General Relativity is not renormalizable and therefore not applicable for very small distances, such as those associated with the beginning of the universe, central point of study at the BCGT. On the other hand, HLGT, due to its anisotropic space-time scaling, is not Lorentz invariant in the high energy UV regime. However, for small distances, the incorporation of higher order derived terms in the spatial components of the curvature to the usual Einstein–Hilbert action, gives rise to a theory free of ghosts and, therefore, HLGT is more appropriate to describe quantum effects of the gravitational field, as for instance vacuum decay processes in the early stages of the universe [24]. The parameters of the theory are the critical exponent z and the foliation parameter λ, associated with a restricted foliation compatible with the Lifshitz scaling. In the low energy limit $z \to 1$ the Lorentz invariance is recovered. In the infrared limit, to recover the full diffeomorphisms symmetry and the usual foliation of the ADM formalism, the $z \to 1$ limit must be accompanied by the limit $\lambda \to 1$ [24].

The WdW equation solutions, represented by a geometric functional of compact manifolds and matter fields, describe the evolution of the quantum wave function of the Universe [25,26]. A puzzling aspect of the WdW equation however is the absence of the time variable. According to [22], the main problem with this issue in quantum gravity is perhaps its closeness to a classical space–time picture. For Rovelli [27] the absence of time is a feature of the classical Hamilton–Jacobi formulation of general relativity, and the wave function is only a function of the "3-geometry", namely the equivalence class of metrics under a diffeomorphism, and not of the specific coordinate dependent form of the metric tensor. According to the second law of thermodynamics, forward in time represents the direction in which entropy increases and in which we obtain information, so the flow of time would represent a subjective feature of the universe, not an objective part of physical reality [27]. In this realm, in which the observable universe does not exhibit time-reversal symmetry, events, rather than particles or fields, are the basic constituents of the universe, implying that the evolution of physical quantities is related to the description of the relationship between events [27–30]. For instance, given the wave function of the universe as a functional constrained to a region configuration of a super-space that contains a three-surface and matter fields, represented by Φ, where the metric is described by h_{ij} the corresponding Wdw wave function $\Psi(h_{ij}, \Phi)$ may be interpreted as describing the evolution of $\Psi(\Phi)$ in the physical variable Φ.

In this contribution we go beyond the previous formulation. The momentum operators are deduced and the quantum version of the Hamiltonian is obtained by addressing the well-known ambiguity on the ordering of operators in the Wheeler–DeWitt Hamiltonian[6].

Although there exists infinite possibilities, a parameter α which defines the ordering of the operators was restricted, for comparison purposes, to a special class of values, following the options of some authors. More precisely, $\alpha = 1$ [32], and $\alpha = 0, 2$ [33].

Determining the composition of matter and energy in the Universe represents one of the most important challenges in cosmology. The most recent developments suggest that the Universe's content, other than dark matter, is unaccounted for or missing. In this contribution, in addition to a branched universe filled with underlying background vacuum energy, primordial matter and radiation, in order to make contact with standard model calculations, we supplement this formulation with baryon matter, dark matter and quintessence contributions. Quintessence—a time-varying, spatially inhomogeneous, and negative pressure component of the cosmic fluid—is a dynamic ingredient: its energy density and pressure vary with time and is spatially inhomogeneous [34,35]. The main motivation to consider the presence of quintessence is to address, in the future, the so-called "coincidence problem", related to the initial conditions necessary to produce the quasi-coincidence of the densities of matter and quintessence in the present stage of the universe [34,35]. Furthermore, in the approaches commonly presented in the literature, the material composition of the primordial universe refers to the plasma of quarks and gluons and leptons, and with regard to dark matter, a frequent approach is that of a geometric effect through a cosmological constant. In this work, aiming at the future study of the matter–antimatter asymmetry of the universe and baryogenesis, as well as the dark matter described by a kind of cosmic fluid, with an equation of state of the form $P = \omega\rho$, we consider the contribution of the additional terms. The theory coupling parameters, $g_i (i = 0, 1, ..., 9)$, are in turn dimensionless running couplings constants.

Finally, the boundary conditions for the wave function of the Universe are imposed by assuming the Bekenstein criterion, which indicates the existence of an universal upper bound of magnitude $2\pi R / \hbar c$ to the entropy-to-energy ratio S/E of an arbitrary system of effective radius R.

We proceed as follows:

- The line element squared within the branched cosmology is defined and can be retrieved in Refs. [9,13].
- The action is defined, using the Horǎva–Lifshitz theory of gravity, which is the General Relativity augmented by counter-terms to render the theory regularized. For more information, please consult Ref. [24]. The basic ingredients are now expressed in terms of $\ln[\beta(t)]$, which substitutes the standard scale factor $a(t)$. In Section 2.1, the classical impulse variable is defined and the classical Hamiltonian constructed.
- A quantization procedure is applied, elevating the momentum operator and Hamiltonian to operators. As a result we obtain the Wheeler–DeWitt equation.
- Following this path, a parameter α appears which defines the ordering of the operators, as applied in the past to the Wheeler–DeWitt equation. This leaves us with three possible equations.
- These equations are solved using the Range–Kutta numerical analysis iterative method. Unlike the approaches usually found in the literature, in our calculations we do not use approximations. We then obtain new analytic solutions, depending on the boundary conditions based on the Bekenstein's theorem, which provides an upper limit for the entropy. For more information, please consult [10–13,36].

2. Extended Class of the Branched Quantum Cosmological Solutions

In what follows we investigate a branched quantum formulation of the WDW equation, whose the only dynamical variable, the helix-like scale factor analytically continued to the complex plane, as well as its corresponding conjugate momentum, are raised to the rank of quantum operators.

The equation developed by Wheeler and DeWitt, in 1967, represents a fundamental approach for describing quantum gravity [21]. As stressed before, this model, based on the Arnowitt-Deser-Misner decomposition of canonical general relativity in $3 + 1$ dimensions, is additionally complemented by a boundary term proposed by the authors of Refs. [37–40]. Dirac's canonical quantization procedure applied to the Einstein–Hilbert action results in a second-order functional differential equation defined in a configuration superspace, whose solutions depend in general on a three-dimensional induced metric and matter fields [21,25,26,40].

2.1. Branch-Cut Formulation of the Weeler-DeWitt Equation

The complex scale factor $\ln^{-1}[\beta(t)]$ represents, in branched cosmology, as stressed before, the only dynamical variable[7]. The branched manifold $\mathcal{M}$ is in turn layered on hypersurfaces, Σ_t, which are restricted to Riemann foliation leafs, characterized by a complex time parameter, t, with the normalized branching line element analytically continued in 4 dimensions defined as [8,9]

$$ds_{[ac]}^2 = -\sigma^2 N^2(t)c^2 dt^2 + \sigma^2 \left(\ln^{-1}[\beta(t)]\right)^2 \left[\frac{dr^2}{(1 - kr^2(t))} + r^2(t)\left(d\theta^2 + \sin^2\theta d\phi^2\right)\right]. \quad (3)$$

In expression (3), the variables r and t represent, respectively, real and complex spacetime parameters and k the spatial curvature of the multiverse, more specifically, negatively curved (k = −1), flat (k = 0) or positively curved (k = 1) spatial hypersurfaces. $N(t)$ in turn represents the lapse[8] function with $\sigma^2 = 2/3\pi$ denoting a normalisation factor.

In what follows, we consider as a starting point the renormalizable Hořava–Lifshitz theory of gravity whose action, given by $\mathcal{S}_{HL}$, employs terms dependent on the scalar curvature of the Universe and its derivatives, in different orders, defined in the form [20,41]:

$$\begin{aligned}
\mathcal{S}_{HL} = {} & \frac{M_P}{2} \int d^3x dt N \sqrt{-g} \Big\{ K_{ij}K^{ij} - \lambda K^2 - g_0 M_p^2 - g_1 R - g_2 M_P^{-2} R^2 - g_3 M^{-2} R_{ij} R^{ij} \\
& - g_4 M_P^{-4} R^3 - g_5 M^{-4} R(R_j^i R_i^j) - g_6 M^{-4} R_j^i R_k^j R_i^k - g_7 M_P^{-4} R \nabla^2 R \\
& - g_8 M_P^{-4} \nabla_i R_{jk} \nabla^i R^{jk} \Big\};
\end{aligned} \quad (4)$$

in this expression as previously informed g_i denotes the running coupling constants associated to the curvature-dependent terms and its derivatives, M_P represents the Planck mass, and ∇_i are the covariant derivatives. The branching Ricci components of the three dimensional metrics in Equation (4) are determined by imposing a maximum symmetric surface foliation [13]. We then obtain

$$R_{ij} = \frac{2}{\sigma^2 \ln^{-2}[\beta(t)]} g_{ij}, \quad \text{and} \quad R = \frac{6}{\sigma^2 \ln^{-2}[\beta(t)]}, \quad (5)$$

where R represents the branching scalar curvature. The trace of the extrinsic curvature tensor, K_{ij}, which measures geometry modifications as well as the deformation rates of the normal to a hypersurface as it is transported from one point to another, corresponds to a sub-manifold, which depends on the particular embedding and takes the form (for the details see [13])

$$K = K^{ij} g_{ij} = -\frac{3}{2\sigma N} \frac{\left(\frac{d}{dt}\ln^{-1}[\beta(t)]\right)}{\ln^{-1}[\beta(t)]}. \quad (6)$$

Through the use of standard canonical procedures of quantum field theory, a Lagrangian density and the Hamiltonian of the model can be obtained (see [13,20,24,33,41–44]).

3. Spacetime Topological Canonical Quantization

The Lagrangian density of the model is quantized, trough a spacetime topological canonical quantisation[9], by raising the Hamiltonian, the helix-like complex scale factor of the branched gravitation as well as the corresponding conjugate momentum to the category of quantum operators. The resulting formulation describes the evolution of the wave function of the Universe—associated with hyper-surfaces $\Sigma_{\ln}$ analytically continued to the complex plane—in the cosmic scale factor $\ln^{-1}[\beta(t)]$.

Changing variable in the form $u(t) \equiv \ln^{-1}[\beta(t)]$, with $du \equiv d\ln^{-1}[\beta(t)]$, the conjugate momentum p_u of the original branching gravitation dynamical variable $\ln^{-1}[\beta(t)]$ becomes

$$p_u = -\frac{u(t)}{N}\frac{du(t)}{dt}. \tag{7}$$

As a result of applying these standard procedures, the following branching Hamiltonian results (for the details see [13,20,24,33,41–44])

$$\mathcal{H} = \frac{1}{2}\frac{N}{u(t)}\left[-p_u^2 + g_k u^2(t) - g_\Lambda u^4(t) - g_r - \frac{g_s}{u^2(t)}\right], \tag{8}$$

with the dimensionless running coupling constants redefined as [41,42]

$$g_k \equiv \frac{2}{3\lambda - 1}; \quad g_\Lambda \equiv \frac{\Lambda M_{Pl}^{-2}}{18\pi^2(3\lambda - 1)^2}; \quad g_r = 24\pi^2(3g_2 + g_3);$$

$$g_s \equiv 288\pi^4(3\lambda - 1)(9g_4 + 3g_5 + g_6). \tag{9}$$

In these expression, g_k, g_Λ, g_r, and g_s represent, respectively, the curvature, cosmological constant, radiation, and stiff matter coupling constant contributions. The g_r, and g_s coupling constants can be positive or negative, without affecting the stability of the solutions. Stiff matter contribution in turn is determined by the $\rho = p$ condition in the corresponding equation of state.

The quantisation of the Lagrangian density is achieved by raising the Hamiltonian, the new dynamical variable $u(t)$ and the corresponding conjugate momentum p_u to the category of operators, represented, respectively, as $\hat{H}(t)$, $\hat{u}(t)$, and $\hat{p}_u$:

$$\mathcal{H}(t) \to \hat{H}(t); \quad u(t) \to \hat{u}(t); \quad \text{and} \quad p_u \to \hat{p}_u = -i\hbar\frac{\partial}{\partial u(t)}. \tag{10}$$

In what follows, for simplicity, the *hat* symbol is not used in the operators $\hat{u}$ and $\hat{p}_u$ most of the time, as well as in most part of equations the time-dependence on the new variable $u(t)$.

Ambiguities in the ordering of the quantum operators are overcome with the introduction of a set of ordering factors, given by $\alpha = [0, 1, 2]$, following options found in the literature [32,33], as previously mentioned, with p^2 defined as

$$p^2 \equiv -\frac{1}{u^\alpha(t)}\frac{\partial}{\partial u(t)}\left(u^\alpha(t)\frac{\partial}{\partial u(t)}\right). \tag{11}$$

The approach based on the insertion of a set of ordering factors, makes it possible to obtain a broader class of solutions for the Universe wave function.

Combining (8) and (11), we get the subsequent expression for the Wheeler–DeWitt equation for the wave function of the Universe, $\Psi(t)$:

$$\mathcal{H}(t)\Psi(u) = \left(-\frac{1}{u^\alpha}\frac{d}{du}\left(u^\alpha\frac{d}{du}\right) + V(u)\right)\Psi(u) = 0 \tag{12}$$

with the effective potential[10]

$$V(u) = -\eta_r + \eta_m u + \eta_k u^2 + \eta_q u^3 - \eta_\Lambda u^4 - \frac{\eta_s}{u^2},\tag{13}$$

which we supplemented with two additional terms, $\eta_m u$, that describes the contribution of baryon matter combined with dark matter, and $\eta_q u^3$, a quintessence-term. From this expression, for $\alpha = 0$, we obtain the following equation under the action of a real potential[11] represented by $V(u)$:

$$\left(-\frac{d^2}{du^2} + V(u)\right)\Psi(u) = 0.\tag{14}$$

With the choice $\alpha = 1$ in expression (12), we get the equation

$$\left(-\left\{\frac{1}{u}\frac{d}{du} + \frac{d^2}{du^2}\right\} + V(u)\right)\Psi(u) = 0.\tag{15}$$

Finally, the choice $\alpha = 2$ in expression (12), results in the following equation

$$\left(-\left\{\frac{2}{u}\frac{d}{du} + \frac{d^2}{du^2}\right\} + V(u)\right)\Psi(u) = 0.\tag{16}$$

With a view to comparing results based on the standard formulation, in what follows, we set up the dimensionless coupling parameters of the effective potential with values found in the literature, complementing the coupling constants of baryon and dark matter and quintessence with a parametrization based on the total density parameter, Ω_0, which describes the ratio between the total average density of matter and energy in the early Universe, ρ_T and the critical density, ρ_{crit}. The most accepted value of the density parameter nowadays is:

$$\Omega_0 \equiv \frac{\rho_T}{\rho_{crit}} = \quad \Omega_B + \Omega_{DM} + \Omega_\Lambda \sim 0.04 + 0.23 + 0.73 \sim 1,\tag{17}$$

where Ω_B, Ω_{DM}, and Ω_Λ represent the baryon matter, dark matter and dark energy density parameters, respectively. At this stage of our investigation, we do not intend to obtain numerical data that may support future cosmological observations, but rather to seek first to establish a formal consistency in the treatment of the quantum branch-cut gravitation, with the aim of establishing observational predictions based on a consistent theoretical formulation in the future. There are numerous formulations in the literature, based on standard cosmology, that consistently deal with this problem, using improved technical models. Just to name a few of these, we indicate [13,20,24,27,33,41–44], among many others.

Figures 2–5 show the behavior of the effective potential for sets of values of the running coupling constants. The behavior of the effective potential shows a domain of the stiff matter term, contributing for the presence of singularities, both in the expansion and contraction regions. These results show that the increase of the running coupling constants of the stiff matter produces an enlargement of the singularity domain region. Evidently, a more rigorous analysis of the role and consistency of these parameterizations is necessary. For example, the adoption of coupling constants based on energy density parameter [33], or, in the absence of the stiff matter, to examine the relative contributions of the other contributions. In any case, a lesson learned from this work is the need to seek formal alternatives for the inclusion of such contributions so as not to reinforce, —although such a conclusion is far from categorical—, in an artificial and inconsistent way the dominance of certain alternatives over others.

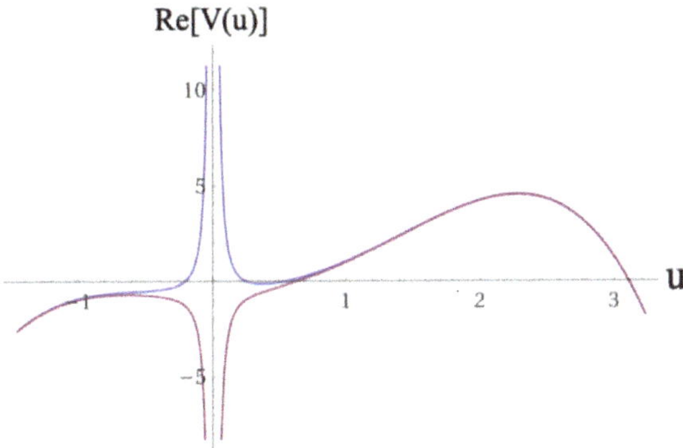

Figure 2. Plot of the real part of the potential defined in Equation (13). In the top figure the coupling constants values are: $\eta_r = 0.6$, $\eta_m = 0.2855$, $\eta_k = 1$, $\eta_q = 0.7$, $\eta_\Lambda = 1/3$, and $\eta_s = -0.03$. In the bottom figure the coupling constants values are: $\eta_r = 0.6$, $\eta_m = 0.2855$, $\eta_k = 1$, $\eta_q = 0.7$, $\eta_\Lambda = 1/3$, and $\eta_s = +0.03$. Values of parameters taken from [43,46,47].

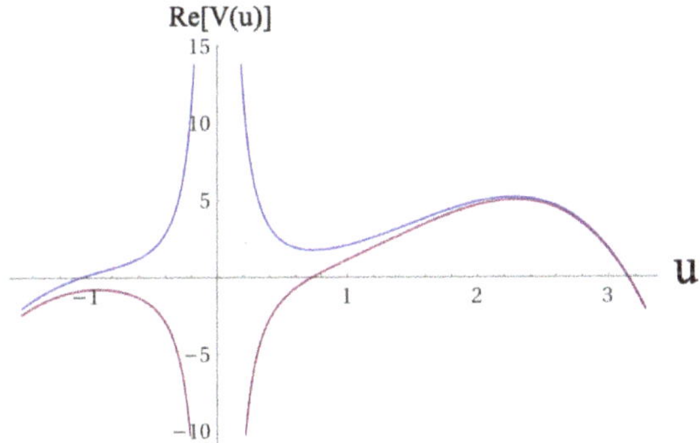

Figure 3. Similar plot of the previous figure. Coupling constants values in the top figure: $\eta_r = 0.024$, $\eta_m = 0.2855$, $\eta_k = 1$, $\eta_q = 0.7$, $\eta_\Lambda = 1/3$, and $\eta_s = -0.468$. Coupling constants values in the bottom figure: $\eta_r = 0.024$, $\eta_m = 0.2855$, $\eta_k = 1$, $\eta_q = 0.7$, $\eta_\Lambda = 1/3$, and $\eta_s = +0.468$. Values of parameters taken from [43,46,47].

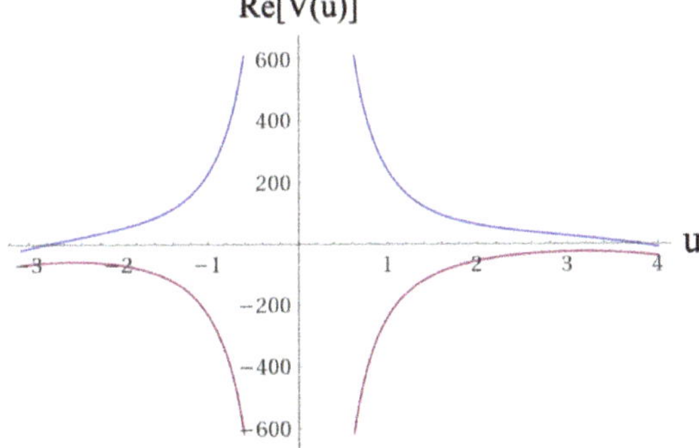

Figure 4. Similar plot of the previous figure. Coupling constants values in the top figure: $\eta_r = 0.0$, $\eta_m = 0.2855$, $\eta_k = 1$, $\eta_q = 0.7$, $\eta_\Lambda = 1/3$, and $\eta_s = -234.0$. Coupling constants values in the bottom figure: $\eta_r = 0.0$, $\eta_m = 0.2855$, $\eta_k = 1$, $\eta_q = 0.7$, $\eta_\Lambda = 1/3$, and $\eta_s = +234.0$. Values of parameters taken from [43,46,47].

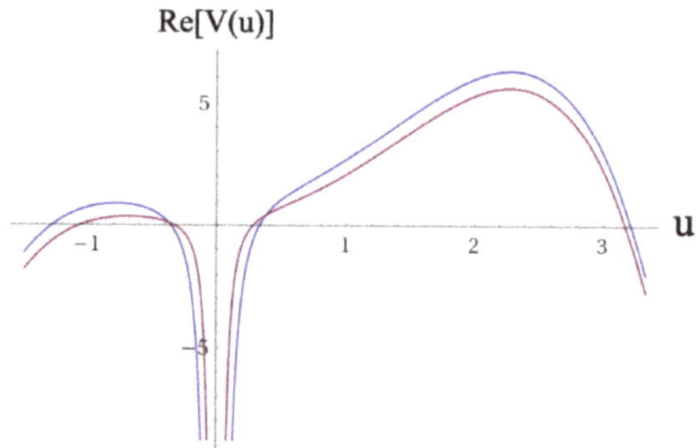

Figure 5. Similar plot of the previous figure. Coupling constants values in the top figure: $\eta_r = -1.22$, $\eta_m = 0.2855$, $\eta_k = 1$, $\eta_q = 0.7$, $\eta_\Lambda = 1/3$, and $\eta_s = 0.15$. Coupling constants values in the bottom figure: $\eta_r = -0.5$, $\eta_m = 0.2855$, $\eta_k = 1$, $\eta_q = 0.7$, $\eta_\Lambda = 1/3$, and $\eta_s = 0.05$. Values of parameters taken from [43,46,47].

3.1. Complex Conjugation of the Friedmann's-Type Wave Equations

In the branching gravitation, the Friedmann's-type equations, analytically continued to the complex plane, and expressed in terms of the new variables $u(t)$, are [7–9]:

$$\left(\frac{\frac{d}{dt}u(t)}{u(t)} \right)^2 = \frac{8\pi G}{3}\rho(t) - \frac{kc^2}{u(t)} + \frac{1}{3}\Lambda, \tag{18}$$

and

$$\left(\frac{\frac{d^2}{dt^2}u(t)}{u(t)} \right) = -\frac{4\pi G}{3}\left(\rho(t) + \frac{3}{c^2}p(t) \right) + \frac{1}{3}\Lambda, \tag{19}$$

where Λ represents the cosmological constant (see Section 1). The corresponding complex conjugated Friedmann's-type equations are:

$$\left(\frac{\frac{d}{dt}u^*(t^*)}{u^*(t^*)} \right)^2 = \frac{8\pi G}{3}\rho^*(t^*) - \frac{kc^2}{u^*(t^*)} + \frac{1}{3}\Lambda^*, \tag{20}$$

and

$$\left(\frac{\frac{d^2}{dt^2}u^*(t^*)}{u^*(t^*)} \right) = -\frac{4\pi G}{3}\left(\rho^*(t^*) + \frac{3}{c^2}p^*(t^*) \right) + \frac{1}{3}\Lambda^*. \tag{21}$$

Equations (18)–(21) underlie the scenarios of branched gravitation in the imaginary sector, as discussed before (see Figure 6): in the first scenario, in the region before the primordial singularity, there is a continuous evolution of the Universe around a branch-cut in the transition region as a function of an imaginary time parameter, conjugated to the corresponding time parameter of the later evolutionary region and no primordial singularity occurs; in the second scenario, the branch-cut and the branch point disappear after realization of the imaginary time by means of a Wick rotation, then this parameter is replaced by the real and continuous thermal time, the temperature. As a result, a parallel evolutionary mirror universe, adjacent to our own, is nested in the fabric of space and time, with its evolutionary process receding into the cosmological sector of negative thermal time. In the following, we adopt, as a consistent formal procedure, conjugated complex versions of expressions (14)–(16). Furthermore, as a consequence of this procedure, solutions of the wave function of the Universe that describe the quantum evolution (in the cosmic scale parameter $\ln^{-1}[\beta(t)]$) of the scenarios described above can be obtained.

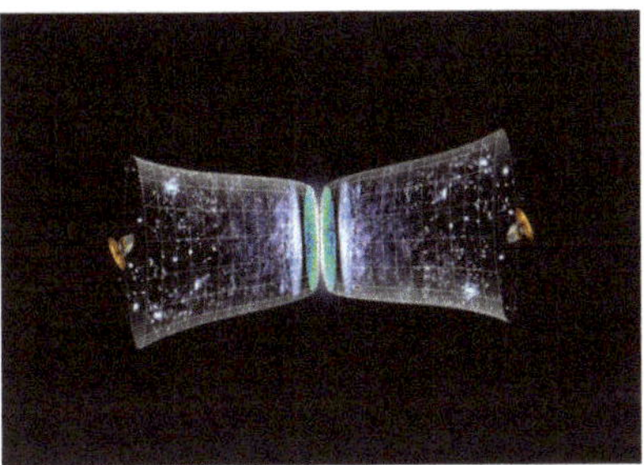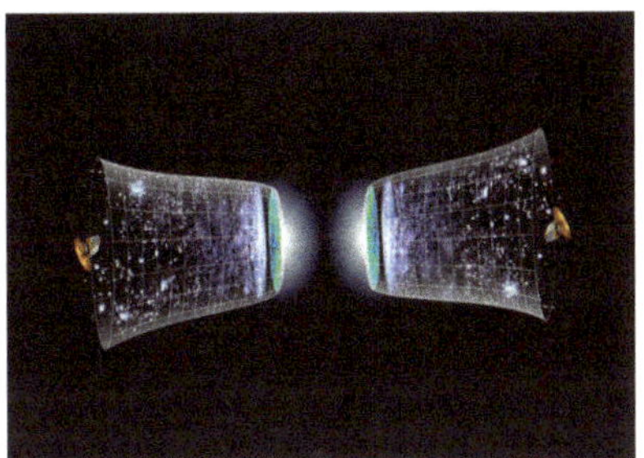

Figure 6. Artistic representations of the cosmic contraction and expansion phases of the branch-cut universe evolution scenarios. On the left figure, the branch-cut universe evolves from negative to positive values of the imaginary cosmological time t_i, circumventing continuously the branch-cut and no primordial singularity occurs, only branch points. On the right figure the branch-cut and branch point disappear after the realisation of imaginary time by means of a Wick rotation, which is replaced here by the real and continuous thermal time (temperature), T. In this scenario, a mirrored parallel evolutionary universe, adjacent to ours, is nested in the structure of space and time, with its evolutionary process going backwards in the cosmological thermal time negative sector. Figures based on artistic impressions [48].

3.2. Solutions and Boundary Conditions

The boundary conditions adopted in this work follows the conventional canons of convergence, as well as stability and continuity of the solutions of the differential equations. Moreover, as a new topic in this contribution, we analyze the boundary conditions of the wave function of the Universe in the light of the Bekenstein criterion [36].

The impossibility of packing the energy and entropy of the primordial Universe into finite dimensions considering spatially connected regions within the particle horizon of a given observer, locus of the most distant points that can be observed at a specific time t_0 in an event, made Bekenstein [36] conjecture an upper bound, given by $\frac{2\pi R}{\hbar c}$, for the entropy S and energy E of a system contained in a spherical region of radius R:

$$\frac{2\pi R}{\hbar c} \geq S/E \quad \text{so} \quad S \leq S_B = \frac{2\pi}{\hbar c} ER, \tag{22}$$

in which S_B denotes the upper limit of Bekenstein entropy.

Considering in a simplified way the proper distance $d(t)$ of a pair of objects, in an arbitrary time t and its relationship with the proper distance $d(t_0)$ in a reference time t_0, $d(t) = u(t)d(t_0)$, this implies that for $t = t_0$, $u(t_0) = 1$. We consider the boundary condition $|u(t_0) = 1|$, assuming the time t_0 as the locus of the most distant points that can be observed, in tune with the Bekenstein criterion. With this assumption, due to the structural characteristics of the proposed effective potential and the extended class of solutions for the wave equations, the wave function of the Universe obeys the following boundary conditions in the expansion sector of the primordial Universe: $\Psi(1) = 1, \Psi'(1) = 0$ and $\Psi(1) = 0, \Psi'(1) = 1$. Similarly, in the contraction sector of the primordial Universe, we have the boundary conditions: $\Psi(-1) = -1, \Psi'(-1) = 0$ and $\Psi(-1) = 0, \Psi'(-1) = -1$, in opposition to the "no boundary" condition [25].

In Figure 7, we plot a sampling solutions family of Equation (14) corresponding to the expansion region of the universe, using a set of values from [43,46,47]. The solutions are in agreement with the corresponding results presented in the literature, although we have not resorted, unlike other authors, to approximations to solve the corresponding differential equations. Approaches adopted by other authors, based on approximations, mainly in the primordial singularity region, limit their numerical analysis, although they have not significantly influenced the global and oscillatory behavior of the solutions.

In Figures 8–13 we show the solutions of Equations (14)–(16). As shown in the figures, for the region domains between $u = -1$ and $u = 1$ the differential equations have no solutions. In our interpretation, this domain corresponds to the region in which a topological quantum leap occurs in accordance with the Bekenstein criterion [11,13].

The main characteristics of these solutions are the oscillatory behavior, whose amplitudes are decreasing as the universe expands, implying an Universe described by oscillating quantum states tending toward a stable ordering at some future time. In the opposite direction, the systematic increase of the oscillatory amplitudes of the wave function as a function of the scale factor $\ln^{-1}[\beta(t)]$ suggests the accumulation of branches, as indicated by the BCGT for restoration of causality. The effect of accumulating branches actually occurs in both the expansion and contraction regions near the transition region.

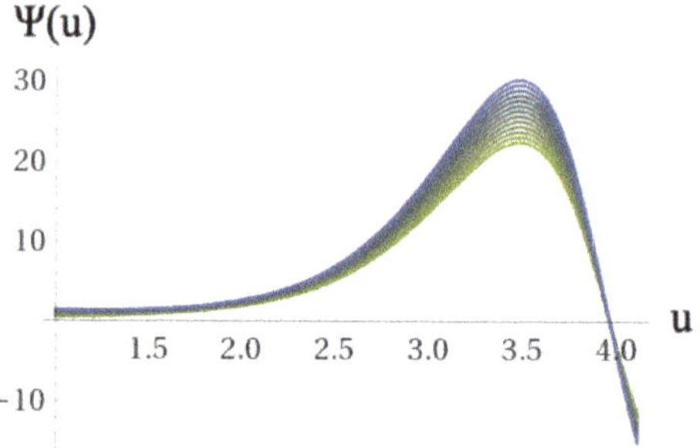

Figure 7. Sampling solution family of Equation (14) with the values of the coupling constants: $\eta_r = 0.6$, $\eta_m = 0.2855$; $\eta_k = 1$; $\eta_q = 0.7$; $\eta_\Lambda = 1/3$; $\eta_s = -0.03$. Values of parameters taken from [43,46,47].

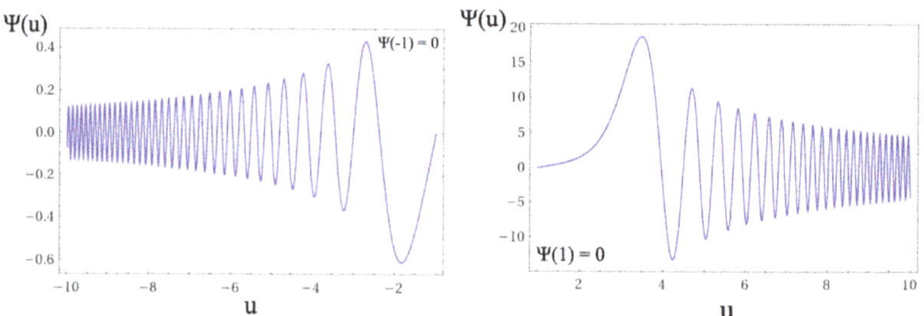

Figure 8. Solutions of Equation (14). The values of the coupling constants are: $\eta_r = 0.6$, $\eta_m = 0.2855$, $\eta_k = 1$, $\eta_q = 0.7$, $\eta_\Lambda = 1/3$, and $\eta_s = -0.03$. Values of parameters taken from [43,46,47].

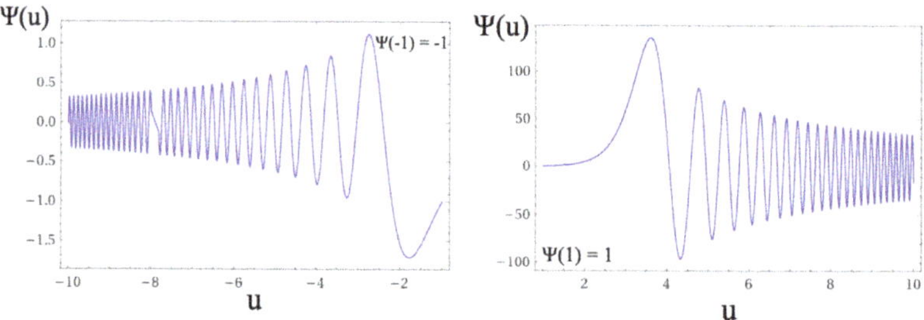

Figure 9. Solutions of Equation (14). The values of the coupling constants are: $\eta_r = -1.22$, $\eta_m = 0.2855$, $\eta_k = 1$, $\eta_q = 0.7$, $\eta_\Lambda = 1/3$, and $\eta_s = 0.15$. Values of parameters taken from [43,46,47].

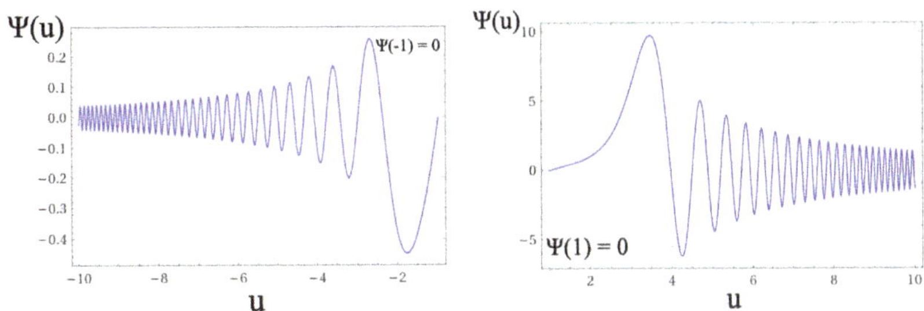

Figure 10. Solutions of Equation (15). The values of the coupling constants are: $\eta_r = 0.6$, $\eta_m = 0.2855$, $\eta_k = 1$, $\eta_q = 0.7$, $\eta_\Lambda = 1/3$, and $\eta_s = -0.03$. Values of parameters taken from [43,46,47].

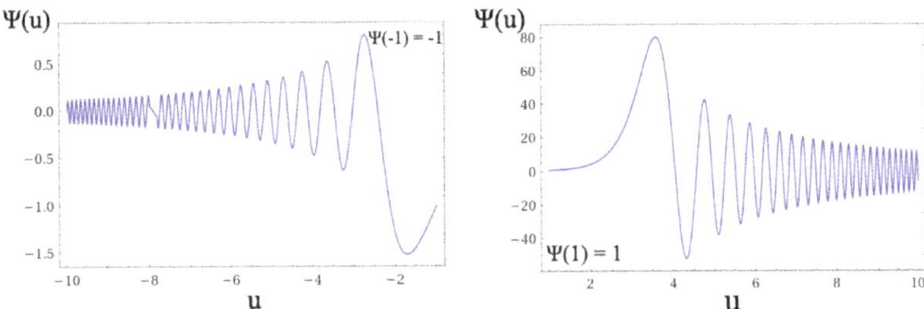

Figure 11. Solutions of Equation (15). The values of the coupling constants are: $\eta_r = 0.6$, $\eta_m = 0.2855$, $\eta_k = 1$, $\eta_q = 0.7$, $\eta_\Lambda = 1/3$, and $\eta_s = -0.03$. Values of parameters taken from [43,46,47].

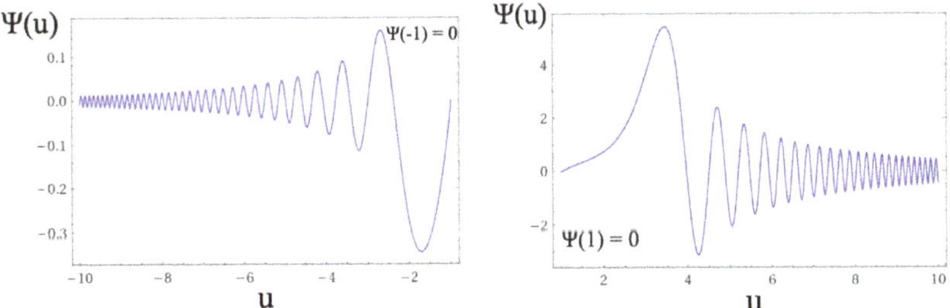

Figure 12. Solutions of Equation (16). The values of the coupling constants are: $\eta_r = 0.6$, $\eta_m = 0.2855$, $\eta_k = 1$, $\eta_q = 0.7$, $\eta_\Lambda = 1/3$, and $\eta_s = -0.03$. Values of parameters taken from [43,46,47].

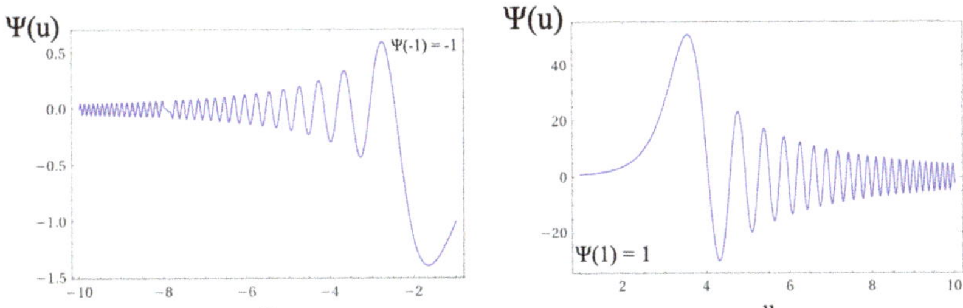

Figure 13. Solutions of Equation (16). The values of the coupling constants are: $\eta_r = 0.6$, $\eta_m = 0.2855$, $\eta_k = 1$, $\eta_q = 0.7$, $\eta_\Lambda = 1/3$, and $\eta_s = -0.03$. Values of parameters taken from [43,46,47].

4. Conclusions

We summarize our most relevant results. We adopt as an underlying proposition a compact universe, filled with homogeneous matter, which exists forever in a quantum state, either static or oscillating, without imposing in an ad hoc way a restriction limit for the cosmological scale factor and for the wave function of the Universe, as its disappearance at any limit of the scale factor (see Ref. [49]). Although its disappearance occurs naturally in the transition region of the branched model, as a natural consequence of the imposition of the Bekenstein criterion, in the expansion and contraction phases, the oscillatory behavior of the wave function of the Universe is characterized by an increase in its amplitudes in the anterior region of the transition phase, indicating consistency with the proposition of accumulating branches to reestablish causality. In the expansion region, going back in time, the same effect occurs. These results indicate that in the limit $u(t) \to \infty$ (or $\ln^{-1}[\beta(t)] \to \infty$), $\Psi(u) \to 0$, implying a Universe described by oscillating quantum states tending towards a stable configuration at some future time. The opposite behavior is verified in the mirror sector of the model. In the mirror sector, the Universe evolves from a stable to an unstable quantum state, and in the visible sector, from an unstable to a stable quantum state. This behavior is contrary to the entropy behavior of the system, which decreases in the evolutionary process of the mirror universe and increases in the visible sector. Making these phenomena compatible seems a challenging task.

Our interpretation of the disappearance of the wave function of the Universe, in turn, in the region between $u = -1$ and $u = 1$, where a topological quantum leap or tunneling occurs according to Bekenstein's criterion, although with a certain harmony with the Vilenkin's quantum tunneling proposal [50], differs from most known proposals for the corresponding boundary conditions[12]. This is because these proposals, although based on different conceptions and assumptions, have in common the prediction of an inflationary stage of evolution in order to reconcile the causality problem of the primordial Universe. In turn, causality involving the horizon size and the patch size, as stressed before, may be accomplished in branch-cut cosmology through the accumulation of branches in the transition region between the present state of the Universe and the past events [10].

The hypothesis of the isotropy of the branch-cut Universe, one of the pillars of cosmology, and its mirror partner may be questioned based on deviations observed in recent decades by means of cosmological probes [52]. Anisotropy of the Universe, in our conception, has two branches to be approached. One branch refers to the evolutionary anisotropy of the mirror universe to our own. Furthermore, another, to the anisotropic directional evolution in both universes. This is a topic that deserves systematic study in the future. Although it is still early for a more effective direction in this study, some aspects deserve attention, such as, for example, the consequences of adopting a non-symmetric approach and a different ordering of the dimensionless thermodynamics connection ϵ, the role of dark matter in the evolution of the branch-cut universe, the role of fluctuations in the primordial spectrum and seeds in the the early universe, and also questions regarding the multiverse content. Likewise, alternative models that address this issue in a complementary way to ours, such as the bouncing model of Ijjas, Steinhardt, and Loeb [17], or Belinsky and Khalatnikov [42,53] proposition for a generic solution of the Einstein equations near their cosmological singularity, based on a generalization of the homogeneous model of Bianchi type IX, deserve our attention in the near future.

The presented proposal strengthens the idea of the transition region of the branched Universe acting as a 'portal' for cosmic material, playing the role this way of an 'eternal seed' [54] for the expanding emergent cosmic scenario.

Finally, a peculiar aspect of the class of solutions presented concerns the insertion of the operators ordering parameter α. As we can see in the presented solutions (Figures 8–13), different values of α, in combination with different choices of running coupling constants affect the amplitudes of the wave function of the Universe and therefore, according to our interpretation, the accumulation of branches in order to restore causality. Evidently, the results presented are still at a preliminary stage of investigation, requiring a more systematic approach in order to broaden its scope.

The conclusions of this work lead to numerous underlying questions, whose understanding has motivated in-progress investigations.

Author Contributions: Conceptualization, C.A.Z.V.; methodology, C.A.Z.V. and B.A.L.B. and P.O.H.B. and J.A.d.F.P. and D.H.; software, C.A.Z.V. and B.A.L.B. and M.R.; validation, C.A.Z.V. and B.A.L.B. and D.H. and P.O.H.B. and J.A.d.F.P.; formal analysis, C.A.Z.V. and B.A.L.B. and P.O.H.B. and J.A.d.F.P. and D.H.; investigation, C.A.Z.V. and B.A.L.B. and P.O.H.B. and J.A.d.F.P. and M.R. and G.A.D.; resources, C.A.Z.V.; data curation, C.A.Z.V. and B.A.L.B.; writing—original draft preparation, C.A.Z.V.; writing—review and editing, C.A.Z.V. and B.A.L.B. and P.O.H.B. and J.A.d.F.P. and D.H. and G.A.D. and M.R.; visualization, C.A.Z.V. and B.A.L.B.; supervision, C.A.Z.V.; project administration, C.A.Z.V.; funding acquisition (no funding acquisition). All authors have read and agreed to the published version of the manuscript.

Funding: This research received no external funding.

Institutional Review Board Statement: Not applicable.

Informed Consent Statement: Not applicable.

Data Availability Statement: Not applicable.

Acknowledgments: P.O.H. acknowledges financial support from PAPIIT-DGAPA (IN100421). The authors wish to thank the referees for suggestions that enabled substantial improvement of this article.

Conflicts of Interest: The authors declare no conflict of interest.

Notes

1. Hawking and Hertog, in 2018, revisited the multiverse concept, conjecturing that the output of eternal inflation does not produce an infinite fractal-type multiverse, but is finite and reasonably smooth.

2. For simplicity the cosmological constant term has been suppressed.

3. We emphasize that these equations do not represent a direct parameterization or generalization of the conventional Friedmann equations described in a single-pole metric and likewise the new cosmic scale factor does not represent a simple parameterization of the standard theory scale factor. Due to the non-linearity of Einstein's equations, such a direct generalization or parametrization would be inconsistent. For the details, see [7–9,12].

4. The impossibility of packing energy and entropy according to the Bekenstein Criterion into a finite size makes the transition phase between contraction and expansion very peculiar, imposing a topology where space-time shapes itself topologically around a branch point.

5. The Hořava–Lifshitz (HL) formulation main goal is to get a renormalizable theory by means of higher spatial-derivative terms of the curvature which are added to the Einstein–Hilbert action [20]. A recurring problem addressed in the analysis of the Hořava–Lifshitz theory of gravity is related to the preservation of general diffeomorphism, a fundamental constraint of general relativity [22]. Although this is not the main topic of discussion, we would like to address that, in the case of restricted foliation preserving diffeomorphism invariance of the Hořava–Lifshitz theory, a well behaved Hamiltonian for gravity may be found [23].

6. For an interesting discussion of this topic see Ref. [31].

7. We emphasize once more that $\ln^{-1}[\beta(t)]$ represents the reciprocal of $\ln[\beta(t)]$ and $\beta(t)$ identifies the range and cuts of the helix-like cosmological factor in branched gravity. $\ln^{-1}[\beta(t)]$ characterizes complex topological leafs of singular foliations by means of Riemann surfaces.

8. $N(t)$ does note represent a dynamical quantity; in turn it denotes a pure gauge variable.

9. As is well know, there are several quantization methods, as for instance, the canonical quantization and the related Dirac scheme, Segal and Borel quantizations, geometric quantization, various ramifications of deformation quantization, Berezin and Berezin–Toeplitz quantizations, prime quantization and coherent state quantization. For a broad overview see [45]. The advantage of the canonical procedure to quantize a classical theory resides in the preservation of the original formal structure, symmetries and conservation laws. The denomination 'spacetime topological canonical quantization' is due to the combination of the conventional canonical quantization procedure applied to a variable, the helix-like complex cosmic scale factor of the branched gravitation, $u = \ln^{-1}[\beta(t)]$, raised to the category of quantum operator, which presents an intricate topology.

10. The condition $\mathcal{H}\Psi(t) = 0$ excludes the multiplicative term $\frac{1}{2}\frac{N}{u(t)}$ in Equation (8).

11 Despite that we consider only the real part of the effective potential, the variable u is complex, and the solutions still have a broader scope, describing the behavior of the wave function of the Universe both for the contraction region, prior to the primordial singularity, and for the later expansion cosmological region.

12 The tunneling boundary condition of Vilenkin [51] in particular has two degrees of freedom: the scale factor and a homogeneous scalar field. A tunneling wave function then describes an ensemble of universes tunneling from "nothing" to a de Sitter space, and then evolving along the lines of an inflationary scenario and eventually collapsing to a singularity [51].

References

1. Manders, K. Domain Extension and the Philosophy of Mathematics. *J. Philos.* **1989**, *86*, 553–562. [CrossRef]
2. Dirac, P.A.M. Complex Variables in Quantum Mechanics. *Proc. R. Soc. A* **1937**, *160*, 48–59.
3. Hess, P.O.; Greiner, W. Pseudo-complex General Relativity. *Int. J. Mod. Phys. E* **2009**, *18*, 51–77. [CrossRef]
4. Hess, P.O.; Schäfer, M.; Greiner, W. *Pseudo-Complex General Relativity*; Springer International Publishing: Cham, Switzerland, 2016.
5. Hess, P.O.; Greiner, W. Pseudo-Complex General Relativity: Theory and Observational Consequences. In *Centennial of General Relativity: A Celebration*; Zen Vasconcellos, C.A., Ed.; World Scientific Publishing Co.: Singapore, 2017; p. 97.
6. Einstein, A. Die Grundlage der Allgemeinen Relativitätstheorie. *Ann. Phys.* **1916**, *49*, 769–822. [CrossRef]
7. Zen Vasconcellos, C.A.; Hadjimichef, D.; Razeira, M.; Volkmer, G.; Bodmann, B. Pushing the Limits of General Relativity Beyond the Big Bang Singularity. *Astron. Nachr.* **2019**, *340*, 857–865. [CrossRef]
8. Zen Vasconcellos, C.A.; Hess, P.O.; Hadjimichef, D.; Bodmann, B.; Razeira, M.; Volkmer, G.L. Pushing the limits of time beyond the Big Bang singularity: The branch cut universe. *Astron. Nachr.* **2021**, *342*, 765–775. [CrossRef]
9. Zen Vasconcellos, C.A.; Hess, P.O.; Hadjimichef, D.; Bodmann, B.; Razeira, M.; Volkmer, G.L. Pushing the limits of time beyond the Big Bang singularity: Scenarios for the branch cut universe. *Astron. Nachr.* **2021**, *342*, 776–787. [CrossRef]
10. Bodmann, B.; Zen Vasconcellos, C.A.; de Freitas Pacheco, J.; Hess, P.O.; Hadjimichef, D. Causality and the arrow of time in the branch-cut cosmology. *Astron. Nachr.* **2022**, *344*, e220086. [CrossRef]
11. de Freitas Pacheco, J.; Zen Vasconcellos, C.A.; Hess, P.O.; Hadjimichef, D.; Bodmann, B. Branch-cut cosmology and the Bekenstein Criterion. *Astron. Nachr.* **2022**, *344*, e220070. [CrossRef]
12. Zen Vasconcellos, C.A.; Hess, P.O.; de Freitas Pacheco, J.; Hadjimichef, D.; Bodmann, B. The branch-cut cosmology: Evidences and open questions. *Astron. Nachr.* **2022**, e20220079. [CrossRef]
13. Hess, P.O.; Zen Vasconcellos, C.A.; de Freitas Pacheco, J.; Hadjimichef, D.; Bodmann, B. The branch-cut cosmology: A topological canonical quantum-mechanics approach. *Astron. Nachr.* **2022**, e20220101. [CrossRef]
14. Aharonov, Y.; Bohm, D. Significance of electromagnetic potentials in the quantum theory. *Phys. Rev.* **1959**, *115*, 485–491. [CrossRef]
15. Wu, K.-D.; Kondra, T.V.; Rana, S.; Scandolo, C.M.; Xiang, G.-Y.; Li, C.-F.; Guo, G.-C.; Streltsov, A. Operational resource theory of imaginarity. *Phys. Rev. Lett.* **2021**, *126*, 090401. [CrossRef]
16. Hawking, S.W.; Hertog, T. A smooth exit from eternal inflation? *J. High Energ. Phys.* **2018**, *4*, 1–17. [CrossRef]
17. Ijjas, A.; Steinhardt, P.J.; Loeb, A. Scale-free primordial cosmology. *Phys. Rev.* **2014**, *D89*, 023525. [CrossRef]
18. Ijjas, A.; Steinhardt, P.J. Bouncing cosmology made simple. *Class. Quantum Gravity* **2018**, *35*, 135004. [CrossRef]
19. Ijjas, A.; Steinhardt, P.J. A new kind of cyclic universe. *Phys. Lett. B* **2019**, *795*, 666–672. [CrossRef]
20. Hořava, P. Quantum gravity at a Lifshitz point. *Phys. Rev. D* **2009**, *79*, 084008. [CrossRef]
21. DeWitt, B.S. Quantum Theory of Gravity. I. The Canonical Theory. *Phys. Rev.* **1967**, *160*, 1113. [CrossRef]
22. Kiefer, C. *Quantum Gravity*, 3rd ed.; Oxford University Press: Oxford, UK, 2012.
23. Klusoň, J. Hamiltonian analysis of nonrelativistic covariant restricted-foliation-preserving diffeomorphism invariant Hořava-Lifshitz gravity. *Phys. Rev. D* **2011**, *83*, 044049. [CrossRef]
24. García-Compeán, H.; Mata-Pacheco, D. Lorentzian Vacuum Transitions in Hořava–Lifshitz Gravity. *Universe* **2022**, *8*, 237. [CrossRef]
25. Hartle, J.B.; Hawking, S.W. Wave function of the Universe. *Phys. Rev. D* **1983**, *28*, 2960. [CrossRef]
26. Hawking, S.W. The boundary conditions of the Universe. *Pontif. Acad. Sci. Scr. Varia* **1982**, *48*, 563.
27. Rovelli, C. The strange equation of quantum gravity. *Class. Quantum Gravity* **2015**, *32*, 124005. [CrossRef]
28. Rovelli, C. *The Order of Time*; Riverhead Books: New York, NY, USA, 2019.
29. Rovelli, C. *Quantum Gravity*; Cambridge University Press: Cambridge, UK, 2004.
30. Rovelli, C.; Smerlak, M. Thermal time and Tolman–Ehrenfest effect: 'temperature as the speed of time'. *Class. Quantum Gravity* **2011**, *28*, 075007. [CrossRef]
31. Steigl, R.; Hinterleitner, F. Factor ordering in standard quantum cosmology. *Class. Quantum Gravity* **2006**, *23*, 3879–3894. [CrossRef]
32. Hawking, S.W.; Page, D.N. Operator ordering and the flatness of the universe. *Nucl. Phys. B* **1986**, *264*, 185. [CrossRef]
33. Vieira, H.S.; Bezerra, V.B.; Muniz, C.R.; Cunha, M.S.; Christiansen, H.R. Class of solutions of the Wheeler–DeWitt equation with ordering parameter. *Phys. Lett. B* **2020**, *809*, 135712. [CrossRef]
34. Caldwell, R.R.; Dave, R.; Steinhardt, P.J. Cosmological Imprint of an Energy Component with General Equation of State. *Phys. Rev. Lett.* **1998**, *80*, 1582. [CrossRef]
35. Zlatev, I.; Wang, L.; Steinhardt, P.J. Quintessence, Cosmic Coincidence, and the Cosmological Constant. *Phys. Rev. Lett.* **1999**, *82*, 896. [CrossRef]
36. Bekenstein, J.D. Universal upper bound on the entropy-to-energy ratio for bounded systems. *Phys. Rev. D* **1981**, *23*, 287. [CrossRef]

37. Gibbons, G.; Hawking, S.W. Cosmological event horizons, thermodynamics, and particle creation. *Phys. Rev. D* **1977**, *15*, 2752–2756. [CrossRef]
38. York, J.W. Boundary terms in the action principles of general relativity. *Found. Phys.* **1986**, *16*, 249–257. [CrossRef]
39. York, J.W. Role of Conformal Three-Geometry in the Dynamics of Gravitation. *Phys. Rev. Lett.* **1972**, *28*, 1082–1085. [CrossRef]
40. Lukasz, A.G. Novel solution of Wheeler–DeWitt theory. *Appl. Math. Phys.* **2014**, *2*, 73–81.
41. Bertolami, O.; Carlos, A.D.; Zarro, C.A.D. Hořava-Lifshitz quantum cosmology. *Phys. Rev. D* **2011**, *84*, 044042. [CrossRef]
42. Maeda, K.-I.; Misonoh, Y.; Kobayashi, T. Oscillating Bianchi IX Universe in Hořava-Lifshitz Gravity. *Phys. Rev. D* **2010**, *82*, 064024. [CrossRef]
43. Cordero, R.; Garcia-Compean, H.; Turrubiates, F.J. Lorentzian vacuum transitions in Hořava–Lifshitz Gravity. *Gen. Relativ. Gravit.* **2019**, *51*, 138. [CrossRef]
44. Garattini, R.; Faizal, M. Cosmological constant from a deformation of the Wheeler–DeWitt equation. *Nucl. Phys. B* **2016**, *905*, 313–326. [CrossRef]
45. Ali, S.T.; Engliš, M. Quantization Methods: A guide for physicists and analysts. *Rev. Math. Phys.* **2005**, *17*, 391–490. [CrossRef]
46. França, U.; Rosenfeld, R. Quintessence models with exponential potentials. *JHEP* **2002**, *0210*, 15. [CrossRef]
47. Hinshaw, G.; Larson, D.; Komatsu, E.; Spergel, D.N.; Bennett, C.L.; Dunkley, J.; Nolta, M.R.; Halpern, M.; Hill, R.S.; Odegard, N. Nine-year Wilkinson Microwave Anisotropy Probe (WMAP) observations: Cosmological parameter results. *Astrophys. J. Suppl. Ser.* **2013**, *208*, 19. [CrossRef]
48. Kornmesser, M. History of the Universe (ESO). 2020. Available online: https://supernova.eso.org/exhibition/1101/ (accessed on 29 May 2023).
49. Damour, T. and Vilenkin, A. Quantum instability of an oscillating universe. *Phys. Rev. D* **2019**, *100*, 083525. [CrossRef]
50. Vilenkin, A. Creation of universes from nothing. *Phys. Lett. B* **1982**, *117*, 25. [CrossRef]
51. Vilenkin, A. Boundary conditions in quantum cosmology. *Phys. Rev. D* **1986**, *33*, 3560. [CrossRef]
52. Migkas, K.; Schellenberger, G.; Reiprich, T.H.; Pacaud, F.; Ramos-Ceja, M.E.; Lovisari, L. Probing cosmic isotropy with a new X-ray galaxy cluster sample through the LX–T scaling relation. *Astron. Astroph.* **2020**, *636*, A15. [CrossRef]
53. Belinskii, V.A.; Khalatnikov, I.M.; Lifshitz, E.M. A general solution of the Einstein equations with a time singularity. *Adv. Phys.* **1982**, *31*, 639. [CrossRef]
54. Mulryne, D.J.; Tavakol, R.; Lidsey, J.E.; Ellis, G.F.R. An Emergent Universe from a loop. *Phys. Rev. D* **2005**, *71*, 123512. [CrossRef]

 *universe*

Essay

Finding My Drumbeat: Applying Lessons Learned from Remo Ruffini to Understanding Astrophysical Transients

Chris Fryer

Center for Theoretical Astrophysics, Los Alamos National Laboratory, Los Alamos, NM 87545, USA; fryer@lanl.gov; Tel.: +1-505-500-2279

Abstract: As with many fields from fashion to politics, science is susceptible to "bandwagon"-driven research where an idea becomes increasingly popular, garnering a growing amount of "scientific" support. Bandwagons allow scientists to converge on a solution, but when the prevailing bandwagon is incorrect or too simple, this rigid mentality makes it very difficult for scientists to find the right track. True scientific innovation often occurs through scientists willing to march to the beat of their own drum. Using examples in the field of astrophysical transients, this paper demonstrates the importance of supporting scientists in their quest to develop their own personal drumbeat.

Keywords: gamma-ray bursts; supernovae; neutron stars; black holes

Citation: Fryer, C. Finding My Drumbeat: Applying Lessons Learned from Remo Ruffini to Understanding Astrophysical Transients. *Universe* **2023**, *9*, 268. https://doi.org/10.3390/universe9060268

Academic Editors: Gregory Vereshchagin, Jorge Armando Rueda Hernández, Narek Sahakyan and Remo Ruffini

Received: 4 May 2023
Revised: 27 May 2023
Accepted: 1 June 2023
Published: 4 June 2023

1. Introduction: Bandwagon Science

The bandwagon effect is a psychological phenomenon in which people perform something primarily because other people are performing it, regardless of their own personal beliefs. This tendency of the alignment of beliefs or viewpoints is also referred to as herd mentality. Although the bandwagon effect was originally used to describe political viewpoints, it occurs in a broad range of situations, including scientific research and funding.

Bandwagon approaches in science can be beneficial to a scientific endeavor. For instance, when new technology or new methods become available (e.g., machine learning, quantum computing, etc.), many funding agencies jump on a bandwagon to fund this research. Funding managers place much of their funds into calls that require this bandwagon science, and scientists who embrace bandwagon science tend to only review highly the science that is on that bandwagon. The strength of this approach is that it directs funding to an up-and-coming topic, jumpstarting its research. The problem with this is that oftentimes this influx of funding is far too great to drive any productive science, and this overfunding one field leaves other fields unfunded. For many funding agencies, the "bandwagon-of-the-month" oscillates rapidly enough that building and retaining scientists to work on the projects is impossible and much of the funding is wasted.

Science fields can also follow bandwagons. These bandwagons generally support a "standard paradigm" explaining some phenomena. The advantage of the bandwagon effect for such standard paradigms is that it focuses research on a specific paradigm. In principle, this research focus will more quickly flesh out the details of a current standard-paradigm. The disadvantage is that this focused research often poorly identifies weaknesses in the paradigm. Because of this, standard-paradigms often remain "standard" long after the evidence clearly shows it is incorrect. In addition, bandwagon science tends to oversimplify models. Typically, progress in these fields occurs because a subset of scientists do not join the bandwagon, preferring to march to the beat of a different drum. In the end, science progresses through a cacophony of drumbeats and, as messy as this approach is, it may be a requirement in the evolution of science.

In this paper, we review a series of examples in astrophysical transients where science progress required scientists willing to move against the bandwagon. Remo Ruffini's approach to science epitomizes this alternative drumbeat and Section 2 discusses how this approach has driven research in gamma-ray bursts (GRBs). His example is an inspiration driving advances and discoveries of supernovae (SNe): Section 3. This paper focuses on these two examples, but bandwagons exist throughout science. We conclude with a broader discussion of the implications for science as a whole.

2. Gamma-Ray Bursts

The field of GRBs has been prone to bandwagon beliefs that often oversimplify the physics. Although these simplifications make it easier to explain a model, they often neglect physics or processes that are crucial in interpreting observations, ultimately delaying progress in the field. In some cases, the standard-paradigm science case was much weaker than the actual science case without the bandwagon simplifications, making it more difficult for the GRB community to justify new instruments and science studies. As such, the history of GRB science provides examples of how a paradigm can dominate a field and the importance of science outside of the standard paradigm.

2.1. Gamma-Ray Emission

GRB emission models and the reluctance of that community to envision solutions beyond the standard mechanism is a prime example of how bandwagon science can damage a field. Figure 1 shows the basic picture of a GRB outflow. In the standard model, synchrotron (and synchrotron self-Compton) emission produces both the prompt and afterglow emission [1,2]. In the prompt phase, this model argues that the chaotic emission is produced by variability in the jet power, presumably caused by instabilities in the accretion disk. The variability produces outbursts of different Lorentz factors. When a faster burst hits a slow burst, particle acceleration in the shock produces nonthermal emission that is Lorentz-boosted into the gamma-rays. In this standard paradigm, the afterglow (post-burst of gamma-rays) emission is caused by synchrotron emission as the shock propagates through circumstellar medium.

This model has a lot of strengths. First, with a basic synchrotron (or synchrotron plus self-Compton) model incorporating a range of simply-described parameters, the data can be fit fairly well [2]. In its simplest form, this model could be used to probe the jet Lorentz factor from the prompt emission, and, within this paradigm, the afterglow emission can be used to probe the circumstellar medium; however, this model has a number of issues. The parameters needed to explain the GRB prompt emission (Lorentz factor and fraction of power-law electrons) pushed the limits of what can be produced by both engine and plasma kinetics models. The predictions from the afterglow models argued for circumstellar density profiles that contradicted the strong wind models from the standard collapsar progenitor [3]. For the most part, given these difficult-to-explain results, many of the GRB engine and progenitor theorists moved on to other fields, and progress in our understanding of GRBs stalled. Incredible results require powerful arguments and the simplified bandwagon science did not provide these arguments.

Alternative models exist. At early times, some teams argued that external shocks can explain the prompt emission [4–6]. In these models, variability relied on inhomogeneities in the circumstellar medium. Although seemingly ad hoc, these inhomogeneities are what has been predicted by stellar models [7–13]. Because GRB progenitors are believed to have stronger outflows than normal SNe, these inhomogeneities should be most extreme in long-duration GRB progenitors. These features did not fit into the standard paradigm and, often when something does not fit in a paradigm, features are thrown out with the noise. Thus, even though this signal should be stronger in GRBs than SNe, the first evidence of these inhomogeneities came from early-time emission and shock breakout in supernovae [14–16]. Strict adherence to the GRB standard model prevented GRB observers from noticing evidence of this important aspect of GRB progenitors. The science carried out with GRBs was limited by the dominance of GRB bandwagon scientists.

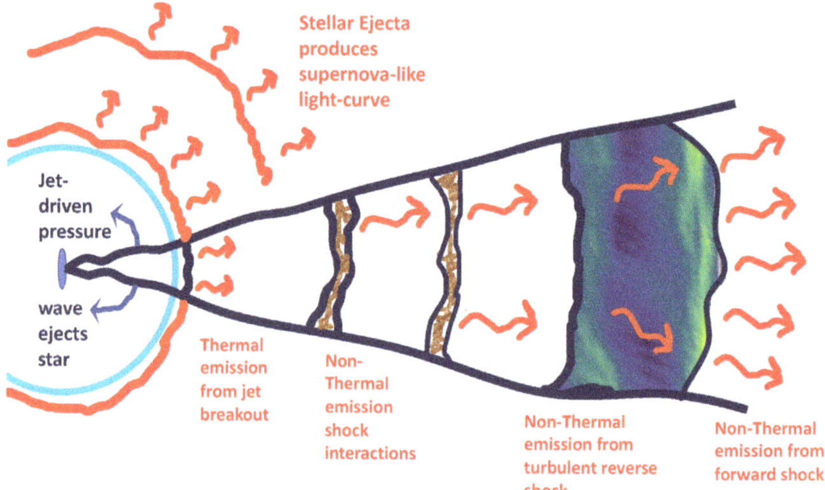

Figure 1. Diagram summarizing many of the emission mechanisms in a GRB-event from a massive star. The standard paradigm assumes that nonthermal emission caused by shock interactions of different outbursts produced in an unstable accretion-driven engine produces the prompt emission and the shock interactions of the jet with the surrounding medium produce the afterglow. However, other sources exist, e.g., nonthermal emission from turbulent particle acceleration in the reverse shock, and thermal emission from the breakout of the shock from the star. The supernova-like explosion that occurs with these explosions occurs because of both the fact that a pressure wave wraps around the star and the disk drives a wind, ejecting the stellar material. Although this is not a standard supernova engine, this nonrelativistic ejecta produces supernova-like or hypernova emission.

It is also possible that turbulence in the reverse shock can cause the needed variability, but this has not been studied in detail. In SNe, this turbulent reverse shock region is also a site of particle acceleration: cosmic ray and high-energy neutrino production [17]. If we could disentangle this portion of the emission from the rest, we might be able to better understand the production of cosmic rays and high-energy neutrinos in GRBs.

Another model that received very little attention was the role of thermal emission in gamma-ray bursts. As we have learned with neutron star (NS) mergers, the emission arises from a range of sources including thermal (powered by shock heating, radioactive decay, and possibly additional engines such as magnetars) and nonthermal (both from the jet as discussed above and the remnant) emission. Initially, the GRB community ignored studies suggesting that the emission (or parts of the emission) could be powered by thermal emission [18–20]. In part, because the theory behind the thermal emission is much better understood than the nonthermal emission, it is a powerful probe of the properties of the GRB ejecta. For example, Figure 2 shows the spectra from this ejecta, assuming a distribution of Lorentz factors where the area (as a function of Lorentz factor) of the thermal emission is $\propto \Gamma^\alpha$. Here, we assume that the minimal Lorentz factor is 1 and the maximum Lorentz factor is either 100 or 200 (for more details, see Lesage et al. in preparation). By studying the emission spectra, we can constrain the structure of the GRB jet at a level that has not been carried out to date, but this requires disentangling the thermal and nonthermal components. This work was delayed until scientists rediscovered the concept of a thermal component that produces the gamma-ray signal [21,22]. It is now applied to many of the new GRB observations, e.g., [23,24].

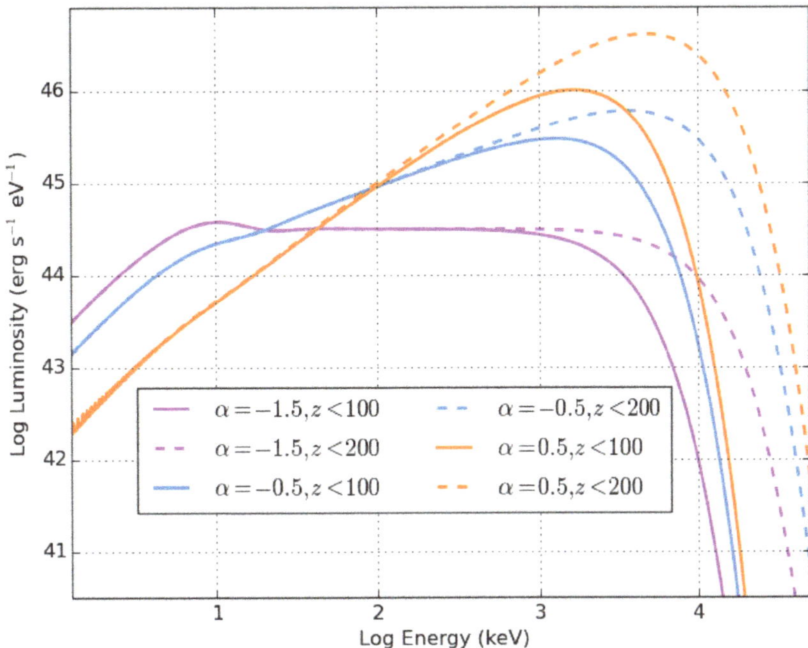

Figure 2. Lorentz-boosted spectra from thermal emission assuming different distributions of the Lorentz factors in the ejecta. The distributions are assume to follow a power law: $\propto \Gamma^{\alpha}$.

Figure 1 also shows emission from the off-jet axis ejecta. The jet sends a pressure wave through the star and this alone has been shown to be able to eject the stellar material [25], but the accretion disk is also likely to blow a strong wind that contributes to the ejection of the stellar material. This mass ejection both powers a supernova-like light curve and, ultimately, limits the accretion, cutting off the engine. Note that a supernova is not required to produce this emission, just the ejection of the stellar material and the pressure wave produced by the jet is sufficient to achieve this.

In this case of GRB observations, by not considering different components to the emission of GRBs, bandwagon scientists delayed progress in our understanding of this emission and limited what we could learn from the data. Many theorists moved on to other fields, believing that nothing new could be learned from the emission itself. In turn, this weakened the science case for further studies, limiting future progress in this field. Pioneers in alternative models have started to revive this field, demonstrating the wealth of information we can gain from the prompt and afterglow emission if we are willing to accept the complex picture of emission mechanisms.

2.2. GRB Engines

The engine or power source for gamma-ray bursts has, for the most part, avoided the trend of bandwagon science to restrict the models studied. A number of broad engine scenarios have been proposed, but two basic engines have been the primary focus in the astrophysical literature: black hole accretion disks (BHADs) and magnetars. The biggest issue with bandwagon beliefs in GRB engines is that scientists have too-easily accepted these models without bearing in mind the properties of each of these models. The refinement of models used the engine properties and weighed the strengths and weaknesses of each model with respect to available data. As the data, and our physical understanding of the engine, evolve, we must reassess each model.

Let us review some of the model properties to better understand this concept (summary in Table 1). Black hole accretion disk models were initially highlighted because they matched (and still match) many of the observed properties of GRBs and it was easier to identify progenitors for these models satisfying the angular momentum requirements (for a review, see [26]). It can produce the observed power (with sufficient beaming) and, because the jet can avoid the accretion disk wind, the model can avoid baryonic contamination and produce high Lorentz-factor flows. The power of this engine depends upon the energy in the disk. Once the mass reservoir in the disk is depleted (no more accretion), the energy in the disk disappears and the engine turns off (limiting the GRB duration). The energy in the disk also decreases with the radius (and hence mass) of the black hole. Collapsar and other massive star disks are fueled by the continued accretion supply from the infalling star, producing long bursts. The compact disks from neutron star mergers accrete quickly and the engine will turn off quickly, producing short bursts only. This simple picture helps explain the different burst-duration populations, but now the data appear to be suggesting a different picture. For example, some GRBs that show evidence of a neuron star merger are occurring in long bursts [27–29].

Table 1. Engine weaknesses and strengths.

Engine	Power	Lorentz Factor	Duration	Formation
BHAD	Yes (depending on beaming)	Yes (disk wind is off-axis)	Duration limited to accretion timescale and black hole size	
Magnetar	Yes (depending on beaming and rotation)	??? (must overcome neutron-star wind)	Duration can extend beyond accretion	Needs high rotation rates; Why does accretion not bury magnetic fields?
NSAD	Yes (simiar to BHAD)	??? (must overcome neutron-star wind)	Duration limited to accretion timescale (NS collapse features?)	Needs high rotation rates

The magnetar engine's strength lies in its ability to explain this long-lived emission (or plateau phase) in neutron-star merger GRBs, but this engine has a number of difficulties. It typically invokes the same progenitor scenarios where accretion is occurring at the same time as the engine is produced, but we know from pulsar studies that this accretion will bury the magnetic fields [30]. The hot neutron star in this scenario also drives a wind that the magnetic engine must plow through, and this propagation is likely to sweep up mass, lowering the Lorentz factor of the jet [31]. Finally, obtaining the power requires very high angular momenta which, although easily achievable in neutron star merger progenitors, is harder to achieve with massive-star scenarios.

Neutron star accretion disk (NSAD) engines are very similar to BHAD engines. The energy in the disk around a neutron star is very comparable to a 3 $M_\odot$ black hole, and the energetics are comparable. However, because there is more angular momentum in the outer layers of a massive star, it is more difficult to produce disks for this engine. This model also has the same baryonic contamination problem that the magnetar engine has.

Other engines exist, invoking pair-plasma power source from an induced gravitational collapse [32–35]. Although quite a few papers have studied this engine, this work was limited to a single team. The potential of this engine mandates further study.

Many times, bandwagon science leads to scientists forgetting the reasons for the standard paradigm, causing scientists to hold on to a simple model for too long and making it difficult to determine key observational tests of the models and distinguish between standard models and newly proposed models.

2.3. GRB Progenitors

As we have seen in the GRB emission picture, bandwagon science focuses on a simple picture even if our fundamental physics understanding argues the problem is much more complex. This trend of bandwagon physics is clearly evident in studies of GRB progenitors of the BHAD engine. Although a broad range of BHAD progenitors were discussed in [36], the community focused on just two scenarios: neutron star mergers for short-duration bursts and collapsars for long-duration bursts. It was argued that the high accretion rates but low accretion timescales for neutron star mergers better fit short–hard bursts and the lower accretion rates with longer disk-feeding timescales of collapsars matched the long–soft bursts [37]. This two-component BHAD paradigm further predicted that long bursts would occur in star-forming regions and short bursts would occur out of these regions and even offset from the host galaxy [36,38].

We already discussed the issues with the BHAD scenario in explaining the long-lived emission (plateau phase) of short bursts, but the simplified collapsar-only progenitor scenario also has difficulties explaining long bursts. First, it remains unclear how the cores of massive stars can collapse with enough angular momentum to form a disk [1]. In addition, it remains true that the supernova-like outbursts accompanying long-duration gamma-ray bursts tend to lack hydrogen and helium features (Type Ic supernovae). The fact that the star had to be a compact (Wolf–Rayet) star was part of the collapsar model. Since the jet in the massive star model must clear out a baryon-free channel, the engine must be able to last at least as long as this clearing phase. Even if a jet is driven in a hydrogen giant star, it will not break out of the star and clear out a low-baryon jet region before the engine turns off. As such, it will not produce large amounts of gamma-ray emission. This is problematic for a giant star, but not for compact helium or carbon/oxygen stars [39]. However, the fact that observations of SNe associated with GRBs are only type Ic SNe argues that only carbon/oxygen stars form GRBs. In the standard collapsar paradigm where the mass-loss occurs through winds, it is difficult to explain why helium stars (which produce type Ib SNe) could not also form GRBs. Wind mass-loss, which tends to remove angular momentum, made it even increasingly difficult to produce black hole accretion disks. Ultra-long bursts are extremely difficult to explain with this engine.

Despite these striking deficiencies, many studies in the field ignore alternative progenitor scenarios. Table 2 shows a subset of the progenitors studied in [36]. Most of the massive star models struggle to explain why the long bursts are mostly associated with supernovae with no helium lines. The exception may be the common envelope formation scenario invoking tidal locking to spin up the core. In this case, sufficient angular momentum may only occur if an extremely tight binary (one consisting of a carbon/oxygen star, a so-called ultra-compact star) is produced. In this scenario, binary interactions both eject mass and spin up the massive star so that it will form a disk when it collapses to a black hole. The common envelope scenario argues that when the massive star in a massive star/compact remnant (NS or BH) binary expands to a giant phase, it envelopes the compact remnant. The subsequent orbital inspiral ejects the envelope (using orbital energy to drive the mass ejection). This can remove the hydrogen envelope and produce a tight binary. If we want to produce a carbon/oxygen star, we must undergo a second common envelope, assuming the helium star also expands into a "giant-like" phase, enveloping the compact remnant. The subsequent common envelope phase would produce an even tighter binary where tidal locking could rapidly spin up the carbon/oxygen star. The problem is that, at this time, only low-mass helium stars undergo giant phases. These low-mass stars are not expected to collapse to form black holes. It might be that this scenario only works for NSAD engines.

Table 2. BHAD progenitors.

Scenario	Duration	Location	Angular Momentum	Associated Transient	Circumstellar Medium
Massive Star					
Wind Mass-Loss	Long bursts	Star-forming regions	Difficult/impossible	Type Ib/Ic	Wind profile
Common Envelope	Long bursts	Star-forming regions	Tidal spin-up	Type Ib/Ic (Tidal Spins could limit to Ic)	Wind plus shell
He–He Merger	Long bursts	Star-forming regions	Difficult	Type Ib/Ic	Wind plus shell
Helium Merger	Long and ultra-long bursts	Star-forming regions and slightly beyond	Can have too much angular momentum	Type Ib/Ic	Wind plus shell
Binaries					
NS/NS	Short bursts	Off-set	$\sim$10 km disk	Disk ejecta only (Kilonova)	Interstellar or intergalactic medium
NS/BH	Short bursts	Off-set	Disk forms for subset	Disk ejecta only (Kilonova)	Interstellar or intergalactic medium
WD/(BH/NS)	Long bursts	Mild off-set	$\sim$10,000 km disk	Fast supernova from disk wind	Mostly interstellar medium

The only massive star progenitor that consistently produces sufficient angular momentum is the helium merger model, where a compact remnant inspirals within a star but merges without ejecting all of the envelope. As it spirals into the center of the massive star, it behaves very similarly to a collapsar accretion scenario. Because of this similarity, this He-merger model was conflated with the collapsar model.

At this time, only the binary-driven hypernova progenitor for the pair-plasma engine has a natural explanation for the hydrogen and helium-poor supernovae associated with GRBs [40]. This alone argues that this engine warrants further study.

Why is it so important to study the exact details of the progenitor instead of relying on the simple bandwagon scenario? First, without identifying the features of each progenitor and its ability to explain GRBs, funding for this research has faltered and little progress has been made in better understanding GRBs, and, in addition, because the different progenitor scenarios have different dependencies with redshift. Many studies assume a given progenitor, e.g., [41] and many of these do not even understand that they have chosen a specific progenitor. Because of this oversimplification in the bandwagon, many of the studies and expectations of metallicity and redshift evolution of GRBs are flawed.

3. Supernovae

GRBs are not the only field in astronomy where a bandwagon focus on simplifying models has led to both misinterpretations of the data and a delay in scientific progress. Studies of supernovae have experienced similar problems, but for many of us who have worked in both fields, lessons learned from the GRB community have helped us overcome issues in the supernova field.

3.1. Thermonuclear Supernovae

The field of thermonuclear supernovae provides a classic example of where bandwagon science not only oversimplified the physics but pushed a paradigm that ultimately is now believed to be just one solution to the problem. For thermonuclear supernovae, the power source is nuclear burning (the conversion of the carbon/oxygen of a white dwarf to silicon and nickel). How this material ignites and burns is not known, but many ideas exist: many-ignition-site deflagration, deflagration transitioning to detonation, detonation of an accretion layer driving the compression of the carbon/oxygen core (e.g., sub-Chandrasekhar models), deflagration in the core igniting a detonation in the accretion layer that then drives a detonation of the core (e.g., gravitationally confined detonation), collisions, double degenerate mergers, etc. (see, for example [42,43]). Although all engines rely on a carbon/oxygen white dwarf, the properties of the white dwarf (e.g., mass), the engine, and the progenitors all vary considerably.

This broad set of models suggests a vibrant field unconstrained by bandwagon science, but this is only after a period of extreme constraints caused by bandwagon science. In the early 1990s, the best light-curve models compared to data strongly supported a Chandrasekhar-mass progenitor for thermonculear supernovae [44,45]. This led the field to primarily consider only engines and progenitors within this paradigm for nearly 10–15 years. Indeed, breaking out of this bandwagon required both the abundance of evidence pushing away from the standard paradigm (including progenitor studies) and a strong-willed push by teams pushing alternative models, e.g., the FLASH team [46].

3.2. Core-Collapse Supernovae

Although the core-collapse supernova (CCSN) community has also homed in on a single paradigm where convection above a collapsed core enhances the conversion of the gravitational potential energy released in the collapse of the core of a massive star into kinetic energy of the explosion [47], this field seems to better allow alternative explosion mechanisms, including jet engines [48]. Indeed, the lack of progress in alternative explosion mechanisms has been more driven by (a) the difficulty in conducting magnetohydrodynamic models and (b) evidence from observations that most stars do not have enough angular momentum for these alternative mechanisms to explain most supernova observations (although such engines probably explain a subset of the observations). This is less an example of bandwagon restrictions and more of scientific limitations. The bigger issues with CCSN lie in interpreting the observations of both shock breakout and later-time light curves.

Shock breakout is the term used to describe the emission produced when the supernova shock produced in the central engine breaks out of a star. While in the star, the radiation is effectively trapped in the flow:

$$v_{\text{radiation}} = v_{\text{diffusion}} \approx \lambda c / D < v_{\text{shock}} \tag{1}$$

where the radiation velocity ($v_{\text{radiation}}$) is well described in the diffusion limit: c is the speed of light, λ is the mean free path, and D is a fraction of the stellar radius. At the high densities of the exploding star, the velocity of the shock (v_{shock}) is much faster than this radiation velocity. As the shock breaks out of the star, the density decreases dramatically, increasing the mean free path and the effective radiation velocity. The radiation quickly changes from trapped in the flow to free-streaming out of the shock. The burst of light from this escape has been observed in a number of events and a simple analytic model was developed to infer the stellar radius [49].

This simple model was exciting for observers because a single observation of the duration and peak luminosity of shock breakout provides direct information about the progenitor star. The problem is that the simplification of the physics means that the the interpretations from the data are simply wrong. The shock breakout signal is affected by asymmetries in the shock [50], asymmetries in the star [51], and asymmetries in the stellar wind [15]. Observations of shock breakout [52] also show that the simple model for shock breakout is incorrect. Unfortunately, this means that the upcoming UltraSAT satellite will teach us less about stars than previously believed. UltraSAT data alone will not be able to disentangle this physics. Detailed models of shock breakout also show the limitations of proposed missions such as STAR-X to perform shock breakout science. In this case, focusing too heavily on the simplified bandwagon model not only delays science, but leads to an inefficient use of scientific funding for new observatories.

Similarly, the light curves of supernovae also have suffered from too-simple models of the emission. Based on state-of-the-art models of light curves, supernova observers estimated the ejecta mass of core-collapse supernovae, arguing that the progenitor masses of these supernovae were all greater that $\sim 20\,M_\odot$ [53]. This directly contradicted models of the supernova engine [54]. In this case, theory pushed forward, ignoring these observed estimates. When direct observations of supernova progenitors [55] agreed with theoretical predictions, little time was lost through the simple paradigm model, but the transient community continues to overinterpret their data. For example, the transient community continues to assume that the peak luminosity of supernovae-like events in supernovae places constraints on the ^{56}Ni yield (assuming that the decay of this radioactive isotope powers the light curve). However, other power sources exist, e.g., shock heating, and there is growing evidence that alternative energy sources must be understood to truly interpret supernova data. Figure 3 shows the luminosity of three supernova explosions, one powered by the decay of radioactive nickel and the other two using a simple shock deceleration model. In these simple models, the light-curve evolution appears very different but, especially for shock heating, a number of affects can alter the evolution. This affect has not been modeled in sufficient detail to determine its importance in SN light curves.

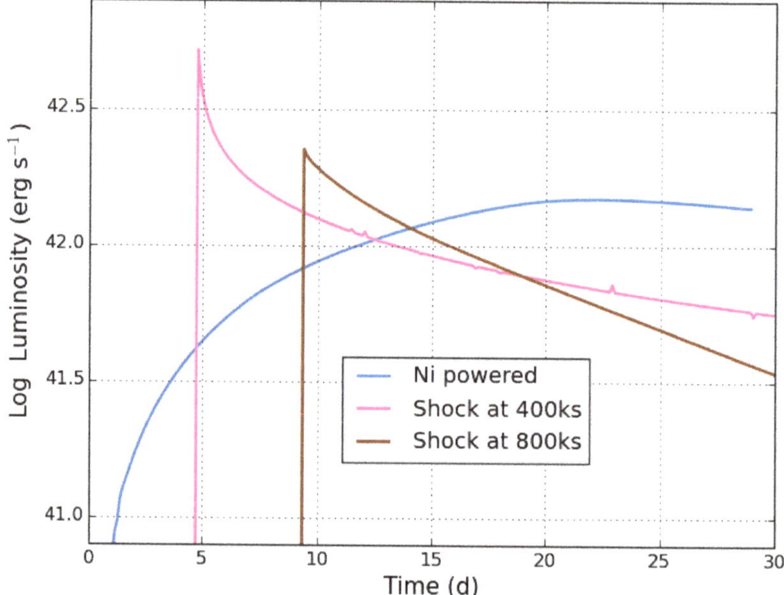

Figure 3. Luminosity for 3 SN explosions with the same Wolf–Rayet progenitor and explosion energy, but one is powered solely by the decay of radioactive nickel (0.05 $M_\odot$ of ^{56}Ni) and the other two are powered by shock heating, where the shock heating model relies on a single shock decelerating the ejecta and converting kinetic energy into thermal energy that drives the emission. The shock heating models differ by having different timescales for the shock deceleration (400 and 800 ks). For more details, see Fryer et al. in preparation.

4. Conclusions

Science, similar to politics, is susceptible to bandwagon beliefs, where an often-simplified model becomes the standard paradigm and the bandwagon scientists are unwilling to consider deviations from this standard paradigm. We discussed examples in astrophysical transients where bandwagon science has limited the science impact, delayed progress, and limited the development of the observations of a specific field, but these examples exist in all areas of research, including work by Gregor Mendel, Ignaz Semmelweis, Alfred Wegener, and George Zweig, to name a few.

Scientific funding is even more susceptible to bandwagon fallacies, often leading to inefficient use of precious research money. Funding managers tend to focus on bandwagon science ideas both because funding managers are enamored by these ideas and because they are easier to pitch to the governments that support their science. In addition, scientists who only conduct bandwagon science tend to be be overly critical of science and science proposals that are not part of existing bandwagons.

An example of such a funding trend may well be the current excitement over machine learning. Advanced statistical methods have their role in science. Funding managers are forcing scientists to apply one such method (machine learning) to their science by diverting their funds to research calls that require this research. Although it is likely that some new discoveries will come from this flux of funding, this overemphasis on a particular tool will also delay progress in many scientific fields.

Similar to Remo Ruffini, scientists must learn to "march to a beat of a different (their own) drum" and encourage others to also do so to ensure rapid scientific progress.

Funding: This work was supported by the US Department of Energy through the Los Alamos National Laboratory. Los Alamos National Laboratory is operated by Triad National Security, LLC, for the National Nuclear Security Administration of U.S. Department of Energy (Contract No. 89233218CNA000001).

Data Availability Statement: The data for the light-curve models shown here will be made available at the LANL Center for Theoretical Astrophysics transient emission site (https://ccsweb.lanl.gov/astro/transient/transients_astro.html, assessed on 5 March 2023) when the full database paper is published.

Acknowledgments: This paper honors Remo Ruffini scientists who have shown how "marching to their own drumbeat" can lead to innovation in science.

Conflicts of Interest: The author declares no conflict of interest.

Abbreviations

The following abbreviations are used in this manuscript:

NS	Neutron Star
NSAD	Neutron Star Accretion Disk
BHAD	Black Hole Accretion Disk
GRB	Gamma-Ray Burst
BH	Black Hole
SN	Supernova
CCSN	Core-Collapse Supernova

Note

[1] It is worth noting that magnetar and NSAD disk engines require even more angular momentum.

References

1. Kobayashi, S.; Piran, T.; Sari, R. Can Internal Shocks Produce the Variability in Gamma-Ray Bursts? *Astrophys. J.* **1997**, *490*, 92. [CrossRef]
2. Sari, R.; Piran, T.; Narayan, R. Spectra and Light Curves of Gamma-Ray Burst Afterglows. *Astrophys. J. Lett.* **1998**, *497*, L17–L20. [CrossRef]
3. Panaitescu, A. An external-shock origin of the relation for gamma-ray bursts. *Mon. Not. R. Astron. Soc.* **2009**, *393*, 1010–1015. [CrossRef]
4. Panaitescu, A.; Mészáros, P. Simulations of Gamma-Ray Bursts from External Shocks: Time Variability and Spectral Correlations. *Astrophys. J.* **1998**, *492*, 683–695. [CrossRef]

5. Chiang, J.; Dermer, C.D. Synchrotron and Synchrotron Self-Compton Emission and the Blast-Wave Model of Gamma-Ray Bursts. *Astrophys. J.* **1999**, *512*, 699–710. [CrossRef]
6. Dermer, C.D.; Mitman, K.E. Short-Timescale Variability in the External Shock Model of Gamma-Ray Bursts. *Astrophys. J. Lett.* **1999**, *513*, L5–L8. [CrossRef]
7. Owocki, S.P.; Rybicki, G.B. Instabilities in line-driven stellar winds. I. Dependence on perturbation wavelength. *Astrophys. J.* **1984**, *284*, 337–350. [CrossRef]
8. Fryer, C.L.; Rockefeller, G.; Young, P.A. The Environments around Long-Duration Gamma-Ray Burst Progenitors. *Astrophys. J.* **2006**, *647*, 1269–1285. [CrossRef]
9. Puls, J.; Vink, J.S.; Najarro, F. Mass loss from hot massive stars. *Astron. Astrophys. Rev.* **2008**, *16*, 209–325. [CrossRef]
10. Herwig, F.; Woodward, P.R.; Lin, P.H.; Knox, M.; Fryer, C. Global Non-spherical Oscillations in Three-dimensional 4π Simulations of the H-ingestion Flash. *Astrophys. J. Lett.* **2014**, *792*, L3. [CrossRef]
11. Quataert, E.; Fernández, R.; Kasen, D.; Klion, H.; Paxton, B. Super-Eddington stellar winds driven by near-surface energy deposition. *Mon. Not. R. Astron. Soc.* **2016**, *458*, 1214–1233. [CrossRef]
12. Jiang, Y.F.; Cantiello, M.; Bildsten, L.; Quataert, E.; Blaes, O.; Stone, J. Outbursts of luminous blue variable stars from variations in the helium opacity. *Nature* **2018**, *561*, 498–501. [CrossRef]
13. Owocki, S.P.; Hirai, R.; Podsiadlowski, P.; Schneider, F.R.N. Hydrodynamical simulations and similarity relations for eruptive mass-loss from massive stars. *Mon. Not. R. Astron. Soc.* **2019**, *485*, 988–1000. [CrossRef]
14. Brown, P.J.; Breeveld, A.A.; Holland, S.; Kuin, P.; Pritchard, T. SOUSA: the Swift Optical/Ultraviolet Supernova Archive. *Astrophys. Space Sci.* **2014**, *354*, 89–96. [CrossRef]
15. Fryer, C.L.; Fontes, C.J.; Warsa, J.S.; Roming, P.W.A.; Coffing, S.X.; Wood, S.R. The Role of Inhomogeneities in Supernova Shock Breakout Emission. *Astrophys. J.* **2020**, *898*, 123. [CrossRef]
16. Bayless, A.J.; Fryer, C.; Brown, P.J.; Young, P.A.; Roming, P.W.A.; Davis, M.; Lechner, T.; Slocum, S.; Echon, J.D.; Froning, C.S. Supernova Shock Breakout/Emergence Detection Predictions for a Wide-field X-Ray Survey. *Astrophys. J.* **2022**, *931*, 15. [CrossRef]
17. Fan, Z.H.; Liu, S.; Wang, J.M.; Fryer, C.L.; Li, H. Stochastic Acceleration in the Western Hot Spot of Pictor A. *Astrophys. J. Lett.* **2008**, *673*, L139. [CrossRef]
18. Ruffini, R.; Bianco, C.L.; Xue, S.S.; Chardonnet, P.; Fraschetti, F.; Gurzadyan, V. On the Instantaneous Spectrum of Gamma-Ray Bursts. *Int. J. Mod. Phys. D* **2004**, *13*, 843–851. [CrossRef]
19. Ruffini, R.; Bianco, C.L.; Xue, S.S.; Chardonnet, P.; Fraschetti, F.; Gurzadyan, V. Emergence of a Filamentary Structure in the Fireball from GRB Spectra. *Int. J. Mod. Phys. D* **2005**, *14*, 97–105. [CrossRef]
20. Bernardini, M.G.; Bianco, C.L.; Chardonnet, P.; Fraschetti, F.; Ruffini, R.; Xue, S.S. Theoretical Interpretation of the Luminosity and Spectral Properties of GRB 031203. *Astrophys. J. Lett.* **2005**, *634*, L29–L32. [CrossRef]
21. Guiriec, S.; Connaughton, V.; Briggs, M.S.; Burgess, M.; Ryde, F.; Daigne, F.; Mészáros, P.; Goldstein, A.; McEnery, J.; Omodei, N.; et al. Detection of a Thermal Spectral Component in the Prompt Emission of GRB 100724B. *Astrophys. J. Lett.* **2011**, *727*, L33. [CrossRef]
22. Guiriec, S.; Mochkovitch, R.; Piran, T.; Daigne, F.; Kouveliotou, C.; Racusin, J.; Gehrels, N.; McEnery, J. GRB 131014A: A Laboratory for Studying the Thermal-like and Non-thermal Emissions in Gamma-Ray Bursts, and the New L^{nTh}_{i}-$E^{nTh,rest}_{peak,i}$ Relation. *Astrophys. J.* **2015**, *814*, 10. [CrossRef]
23. Meng, Y.Z. Evidence of Photosphere Emission Origin for Gamma-Ray Burst Prompt Emission. *Astrophys. J. Suppl.* **2022**, *263*, 39. [CrossRef]
24. Lesage, S.; Veres, P.; Briggs, M.S.; Goldstein, A.; Kocevski, D.; Burns, E.; Wilson-Hodge, C.A.; Bhat, P.N.; Huppenkothen, D.; Fryer, C.L.; et al. Fermi-GBM Discovery of GRB 221009A: An Extraordinarily Bright GRB from Onset to Afterglow. *arXiv* **2023**, arXiv:2303.14172. [CrossRef]
25. MacFadyen, A.I.; Woosley, S.E. Collapsars: Gamma-Ray Bursts and Explosions in "Failed Supernovae". *Astrophys. J.* **1999**, *524*, 262–289. [CrossRef]
26. Fryer, C.L.; Lloyd-Ronning, N.; Wollaeger, R.; Wiggins, B.; Miller, J.; Dolence, J.; Ryan, B.; Fields, C.E. Understanding the engines and progenitors of gamma-ray bursts. *Eur. Phys. J. A* **2019**, *55*, 132. [CrossRef]
27. Rowlinson, A.; O'Brien, P.T.; Metzger, B.D.; Tanvir, N.R.; Levan, A.J. Signatures of magnetar central engines in short GRB light curves. *Mon. Not. R. Astron. Soc.* **2013**, *430*, 1061–1087. [CrossRef]
28. Troja, E.; Fryer, C.L.; O'Connor, B.; Ryan, G.; Dichiara, S.; Kumar, A.; Ito, N.; Gupta, R.; Wollaeger, R.T.; Norris, J.P.; et al. A nearby long gamma-ray burst from a merger of compact objects. *Nature* **2022**, *612*, 228–231. [CrossRef] [PubMed]
29. Rastinejad, J.C.; Gompertz, B.P.; Levan, A.J.; Fong, W.F.; Nicholl, M.; Lamb, G.P.; Malesani, D.B.; Nugent, A.E.; Oates, S.R.; Tanvir, N.R.; et al. A kilonova following a long-duration gamma-ray burst at 350 Mpc. *Nature* **2022**, *612*, 223–227. [CrossRef]
30. Suvorov, A.G.; Melatos, A. Recycled pulsars with multipolar magnetospheres from accretion-induced magnetic burial. *Mon. Not. R. Astron. Soc.* **2020**, *499*, 3243–3254. [CrossRef]
31. Murguia-Berthier, A.; Montes, G.; Ramirez-Ruiz, E.; De Colle, F.; Lee, W.H. Necessary Conditions for Short Gamma-Ray Burst Production in Binary Neutron Star Mergers. *Astrophys. J. Lett.* **2014**, *788*, L8. [CrossRef]

32. Ruffini, R.; Bernardini, M.G.; Bianco, C.L.; Caito, L.; Chardonnet, P.; Dainotti, M.G.; Fraschetti, R.; Guida, R.; Vereshchagin, G.; Xue, S.S. The Role of GRB 031203 in Clarifying the Astrophysical GRB Scenario. In Proceedings of the The 6th INTEGRAL Workshop—The Obscured Universe, Moscow, Russia, 2–8 July 2006; ESA Special Publication: Paris, France, 2007; Volume 622, p. 561. [CrossRef]

33. Ruffini, R.; Bernardini, M.G.; Bianco, C.L.; Caito, L.; Chardonnet, P.; Cherubini, C.; Dainotti, M.G.; Fraschetti, F.; Geralico, A.; Guida, R.; et al. On Gamma-Ray Bursts. In Proceedings of the The Eleventh Marcel Grossmann Meeting On Recent Developments in Theoretical and Experimental General Relativity, Gravitation and Relativistic Field Theories, Berlin, Germany, 23–29 July 2006; pp. 368–505. [CrossRef]

34. Rueda, J.A.; Ruffini, R. On the Induced Gravitational Collapse of a Neutron Star to a Black Hole by a Type Ib/c Supernova. *Astrophys. J. Lett.* **2012**, *758*, L7. [CrossRef]

35. Izzo, L.; Ruffini, R.; Penacchioni, A.V.; Bianco, C.L.; Caito, L.; Chakrabarti, S.K.; Rueda, J.A.; Nandi, A.; Patricelli, B. A double component in GRB 090618: a proto-black hole and a genuinely long gamma-ray burst. *Astron. Astrophys.* **2012**, *543*, A10. [CrossRef]

36. Fryer, C.L.; Woosley, S.E.; Hartmann, D.H. Formation Rates of Black Hole Accretion Disk Gamma-Ray Bursts. *Astrophys. J.* **1999**, *526*, 152–177. [CrossRef]

37. Popham, R.; Woosley, S.E.; Fryer, C. Hyperaccreting Black Holes and Gamma-Ray Bursts. *Astrophys. J.* **1999**, *518*, 356–374. [CrossRef]

38. Bloom, J.S.; Sigurdsson, S.; Pols, O.R. The spatial distribution of coalescing neutron star binaries: implications for gamma-ray bursts. *Mon. Not. R. Astron. Soc.* **1999**, *305*, 763–769. [CrossRef]

39. Fryer, C.L.; Mazzali, P.A.; Prochaska, J.; Cappellaro, E.; Panaitescu, A.; Berger, E.; van Putten, M.; van den Heuvel, E.P.J.; Young, P.; Hungerford, A.; et al. Constraints on Type Ib/c Supernovae and Gamma-Ray Burst Progenitors. *Publ. Astron. Soc. Pac.* **2007**, *119*, 1211–1232. [CrossRef]

40. Fryer, C.L.; Rueda, J.A.; Ruffini, R. Hypercritical Accretion, Induced Gravitational Collapse, and Binary-Driven Hypernovae. *Astrophys. J. Lett.* **2014**, *793*, L36. [CrossRef]

41. Fryer, C.L.; Lien, A.Y.; Fruchter, A.; Ghirlanda, G.; Hartmann, D.; Salvaterra, R.; Upton Sanderbeck, P.R.; Johnson, J.L. Properties of High-redshift Gamma-Ray Bursts. *Astrophys. J.* **2022**, *929*, 111. [CrossRef]

42. Livio, M. The Progenitors of Type Ia Supernovae. In *Type Ia Supernovae, Theory and Cosmology*; Niemeyer, J.C., Truran, J.W., Eds.; Cambridge University Press: Cambridge, UK, 2000; p. 33. [CrossRef]

43. Seitenzahl, I.R.; Townsley, D.M. Nucleosynthesis in Thermonuclear Supernovae. In *Handbook of Supernovae*; Alsabti, A.W., Murdin, P., Eds.; Springer: Cham, Switzerland, 2017; p. 1955. [CrossRef]

44. Hoeflich, P.; Mueller, E.; Khokhlov, A. Light curve models for type IA supernovae - Physical assumptions, their influence and validity. *Astron. Astrophys.* **1993**, *268*, 570–590.

45. Khokhlov, A.; Mueller, E.; Hoeflich, P. Light curves of type IA supernova models with different explosion mechanisms. *Astron. Astrophys.* **1993**, *270*, 223–248.

46. Plewa, T.; Calder, A.C.; Lamb, D.Q. Type Ia Supernova Explosion: Gravitationally Confined Detonation. *Astrophys. J. Lett.* **2004**, *612*, L37–L40. [CrossRef]

47. Herant, M.; Benz, W.; Hix, W.R.; Fryer, C.L.; Colgate, S.A. Inside the Supernova: A Powerful Convective Engine. *Astrophys. J.* **1994**, *435*, 339. [CrossRef]

48. Mösta, P.; Richers, S.; Ott, C.D.; Haas, R.; Piro, A.L.; Boydstun, K.; Abdikamalov, E.; Reisswig, C.; Schnetter, E. Magnetorotational Core-collapse Supernovae in Three Dimensions. *Astrophys. J. Lett.* **2014**, *785*, L29. [CrossRef]

49. Waxman, E.; Katz, B. Shock Breakout Theory. In *Handbook of Supernovae*; Alsabti, A.W., Murdin, P., Eds.; Springer: Cham, Switzerland, 2017; p. 967. [CrossRef]

50. Irwin, C.M.; Linial, I.; Nakar, E.; Piran, T.; Sari, R. Bolometric light curves of aspherical shock breakout. *Mon. Not. R. Astron. Soc.* **2021**, *508*, 5766–5785. [CrossRef]

51. Goldberg, J.A.; Jiang, Y.F.; Bildsten, L. Shock Breakout in Three-dimensional Red Supergiant Envelopes. *Astrophys. J.* **2022**, *933*, 164. [CrossRef]

52. Alp, D.; Larsson, J. Blasts from the Past: Supernova Shock Breakouts among X-Ray Transients in the XMM-Newton Archive. *Astrophys. J.* **2020**, *896*, 39. [CrossRef]

53. Hamuy, M. Observed and Physical Properties of Core-Collapse Supernovae. *Astrophys. J.* **2003**, *582*, 905–914. [CrossRef]

54. Fryer, C.L. Mass Limits For Black Hole Formation. *Astrophys. J.* **1999**, *522*, 413–418. [CrossRef]

55. Smartt, S.J. Progenitors of Core-Collapse Supernovae. *Annu. Rev. Astron. Astrophys.* **2009**, *47*, 63–106. [CrossRef]

Article

Magnetized Black Holes: Interplay between Charge and Rotation

Vladimír Karas [1,*,†] and Zdeněk Stuchlík [2,†]

[1] Astronomical Institute, Czech Academy of Sciences, Boční II 1401, CZ-14100 Prague, Czech Republic
[2] Research Centre of Theoretical Physics and Astrophysics, Institute of Physics, Silesian University in Opava, Bezručovo Nám. 13, CZ-74601 Opava, Czech Republic; zdenek.stuchlik@physics.slu.cz
* Correspondence: vladimir.karas@asu.cas.cz
† These authors contributed equally to this work.

Abstract: Already in the cornerstone works on astrophysical black holes published as early as in the 1970s, Ruffini and collaborators have revealed the potential importance of an intricate interaction between the effects of strong gravitational and electromagnetic fields. Close to the event horizon of the black hole, magnetic and electric lines of force become distorted and dragged even in a purely electro-vacuum system. Moreover, as the plasma effects inevitably arise in any astrophysically realistic environment, particles of different electric charges can separate from each other, become accelerated away from the black hole or accreted onto it, and contribute to the net electric charge of the black hole. From the point of principle, the case of super-strong magnetic fields is of particular interest, as the electromagnetic field can act as a source of gravity and influence spacetime geometry. In a brief celebratory note, we revisit aspects of rotation and charge within the framework of exact (asymptotically non-flat) solutions of mutually coupled Einstein–Maxwell equations that describe magnetized, rotating black holes.

Keywords: black holes; electromagnetic fields; general relativity; microquasars; supermassive black holes

1. Introduction

Classical black holes are described by a small number of parameters; in particular, the mass, electric and magnetic charges, and the angular momentum (spin) [1,2]. As a model of cosmic black holes, these objects are spatially localized and they lack any surface; the resulting spacetime has, by assumption, no material content in the form of fluids that could contribute as a source of the gravitational field. These objects do not support their own magnetic field: just the gravito-magnetic component is induced by rotation [3]. The interacting magnetic field to which astrophysical black holes are embedded is of external origin (Ruffini and Wilson [4]), although it may naturally interact with the Kerr–Newman intrinsic charge [5].

This approach was employed by a number of authors to address the problem of electromagnetic effects near a rotating (Kerr) black hole. On the other hand, self-consistent solutions of coupled Einstein–Maxwell equations for black holes immersed in electromagnetic fields have been studied only within stationary, axially symmetric electro-vacuum models. It soon appeared that the test electromagnetic field approximation was fully adequate for modeling astrophysical sources; however, the long-term evolution of magnetospheres of rotating black holes and the consequences of strong gravity remained still open to further work [6,7]. To explore the latter, the intriguing effects of ultra-strong magnetic fields, we employ an axially symmetric solution that was derived originally in the 1970s in terms of magnetization techniques [7,8].

Although the main aim and the motivation of our present contribution is to briefly summarize some of the aspects of magnetized black holes that have been explored over

Citation: Karas, V.; Stuchlík, Z. Magnetized Black Holes: Interplay between Charge and Rotation. *Universe* **2023**, *9*, 267. https://doi.org/10.3390/universe9060267

Academic Editor: Gonzalo J. Olmo

Received: 1 May 2023
Revised: 25 May 2023
Accepted: 31 May 2023
Published: 3 June 2023

six decades of intensive research, and where the honoree and his collaborators published a number of widely cited discoveries, we will mention also some interesting features of the induced electric charge that occur in this regime and are explored to date. In fact, the generation of magnetic fields goes hand in hand with the creation of corresponding electric fields which always arise in moving media and, for that matter, they appear once a rotating body is involved.

2. Magnetized Kerr–Newman Black Hole in Charge Equilibrium

We can write the system of mutually coupled Einstein–Maxwell equations (Chandrasekhar 1983 [1]),

$$R_{\mu\nu} - \tfrac{1}{2}Rg_{\mu\nu} = 8\pi T_{\mu\nu}, \tag{1}$$

where the source term $T_{\mu\nu}$ is of purely electromagnetic origin,

$$T^{\alpha\beta} \equiv T^{\alpha\beta}_{\text{EMG}} = \frac{1}{4\pi}\left(F^{\alpha\mu}F^{\beta}_{\mu} - \frac{1}{4}F^{\mu\nu}F_{\mu\nu}g^{\alpha\beta} \right), \tag{2}$$

and $^\star F_{\mu\nu} \equiv \tfrac{1}{2}\varepsilon_{\mu\nu}{}^{\rho\sigma}F_{\rho\sigma}$. Let us first consider a strongly magnetized Kerr–Newman (MKN) black hole. This is an electro-vacuum spacetime solution with a regular event horizon that satisfies the conditions of axial symmetry and stationarity. Hence, it adopts a general form [9,10]

$$ds^2 = f^{-1}\left[e^{2\gamma}\left(dz^2 + d\rho^2 \right) + \rho^2\, d\phi^2\right] - f(dt - \omega\, d\phi)^2, \tag{3}$$

with f, ω, and γ being the functions of cylindrical coordinates ρ and z only because of the assumed symmetries. Although in the weak electromagnetic field approximation the Kerr metric gives the line element [11], the case of a strong magnetic field is different, especially at large values of the cylindrical radius. This is because of the magnetic field curving the spacetime and changing its asymptotical characteristics into a non-flat (cosmological) solution (see, e.g., Gal'tsov 1986 [12]).

Christodoulou and Ruffini [13] introduced the magnetic and electric lines of force that are defined, respectively, by the direction of Lorentz force that acts on electric/magnetic charges,

$$\frac{du^\mu}{d\tau} \propto {}^\star F^\mu_\nu\, u^\nu, \qquad \frac{du^\mu}{d\tau} \propto F^\mu_\nu\, u^\nu. \tag{4}$$

In an axially symmetric system, the equation for magnetic lines of force adopts a form that is fully expected on the basis of classical electromagnetism,

$$\frac{dr}{d\theta} = -\frac{F_{\theta\phi}}{F_{r\phi}}, \qquad \frac{dr}{d\phi} = \frac{F_{\theta\phi}}{F_{r\theta}}. \tag{5}$$

By employing the solution generating technique [14], García Díaz 1985 [15] gave a very general and explicit form of the *exact* spacetime metric of a strongly magnetized black hole:

$$ds^2 = |\Lambda|^2\Sigma\left(\Delta^{-1}\,dr^2 + d\theta^2 - \Delta A^{-1}\,dt^2\right) + |\Lambda|^{-2}\Sigma^{-1}A\sin^2\theta(d\phi - \omega\,dt)^2, \tag{6}$$

where $\Sigma(r,\theta) = r^2 + a^2\cos^2\theta$, $\Delta(r) = r^2 - 2Mr + a^2 + e^2$, $A(r,\theta) = (r^2 + a^2)^2 - \Delta a^2\sin^2\theta$ are the well-known metric functions from the Kerr–Newman solution. The event horizon exists for $a^2 + e^2 \le 1$. In the magnetized case, because of the asymptotically non-flat nature of the spacetime, the parameters a and e are not identical with the black hole total spin and electric charge [16]. Moreover, because of the asymptotically non-flat nature of the spacetime, the Komar-type angular momentum and electric charge (as well as the black hole mass) have to be defined by integration over the horizon sphere rather than at radial infinity [17]. The magnetization function $\Lambda = 1 + \beta\Phi - \tfrac{1}{4}\beta^2\mathcal{E}$ is given in terms of the Ernst potentials $\Phi(r,\theta)$ and $\mathcal{E}(r,\theta)$,

$$\Sigma\Phi \;=\; ear\sin^2\theta - \Im e\left(r^2+a^2\right)\cos\theta, \tag{7}$$

$$\begin{aligned}
\Sigma\mathcal{E} \;=\;& -A\sin^2\theta - e^2\left(a^2+r^2\cos^2\theta\right)\\
&+2\Im a\left[\Sigma\left(3-\cos^2\theta\right)+a^2\sin^4\theta - re^2\sin^2\theta\right]\cos\theta.
\end{aligned} \tag{8}$$

The components of the electromagnetic field with respect to orthonormal LNRF components are

$$H_{(r)}+\mathrm{i}E_{(r)} \;=\; A^{-1/2}\sin^{-1}\theta\,\Phi'_{,\theta}, \tag{9}$$

$$H_{(\theta)}+\mathrm{i}E_{(\theta)} \;=\; -(\Delta/A)^{1/2}\sin^{-1}\theta\,\Phi'_{,r}, \tag{10}$$

where $\Phi'(r,\theta)=\Lambda^{-1}\left(\Phi-\tfrac{1}{2}\beta\mathcal{E}\right)$, and the total electric charge Q_H is

$$Q_\mathrm{H} = -|\Lambda_0|^2\,\Im\mathrm{m}\,\Phi'(r_+,0). \tag{11}$$

The magnetic flux $\Phi_\mathrm{m}(\theta)$ across a cap placed in an axisymmetric position on the horizon is then [18]

$$\Phi_\mathrm{m} = 2\pi|\Lambda_0|^2\,\Re\mathrm{e}\,\Phi'\left(r_+,\bar\theta\right)\Big|^{\theta}_{\bar\theta=0}, \tag{12}$$

where $\Lambda_0=\Lambda(\theta=0)$. In Figure 1, the surface plot of the magnetic flux F across the hemisphere $\theta=\pi/2$ is shown as a function of spin parameter a and the electric charge parameter e. The surface on the horizon is defined on the circle $a^2+e^2\le 1$.

The definition interval of the azimuthal coordinate in the magnetized solution needs to be rescaled by a factor Λ_0 (not to be confused with the cosmological term) in order to avoid a conical singularity on the symmetry axis [16], which effectively leads to the increase in the horizon surface area, and thereby also the total magnetic flux threading the event horizon [19]. Let us note that cosmic magnetic fields are limited in strength only by quantum theory effects. In highly magnetized rotators the energy of the magnetic field can be converted into high-energy gamma rays, but such mechanisms require over 10^{12} tesla; we shall not consider this ultra-strong magnetic field in the rest of the paper.

The above-discussed electro-vacuum solutions need to be extended by including an electrically conducting plasma. Once this is introduced into the MKN system, one needs to clarify to what extent the newly emerging role of the Λ term affects the characteristics of the flow of material. This can be investigated in terms of *plasma horizon* and the *guiding centre* approximation, which was originally introduced in the context of accreting black holes by Ruffini [20], Damour et al. [21], and Hanni and Valdarnini [22]. Surfaces of magnetic support were further extended to the case of a black hole that is moving at constant velocity [23,24]. Although these authors considered the case of weak (test) magnetic field in Kerr metric, in a subsequent analysis by Karas and Vokrouhlický [25] we verified that, for astrophysically realistic values of magnetic intensity, the approximate flow lines coincide almost precisely with those constructed for the exact MKN system; they are indistinguishable for practical purposes.

The energy density contained in astrophysically realistic electromagnetic fields turns out to be far too low to influence spacetime noticeably. Test-field solutions are thus adequate for describing weak electromagnetic fields, even those around magnetized neutron stars and cosmic black holes that are currently known.

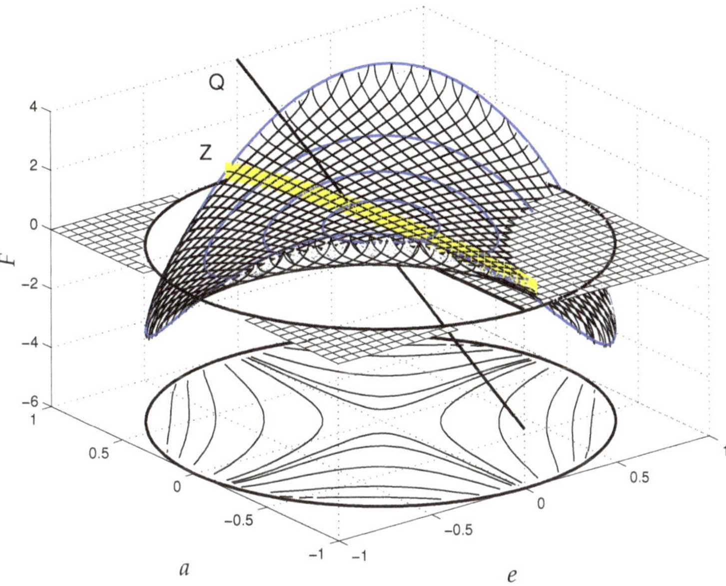

Figure 1. Surface plot of the magnetic flux function, $F(a,e)$, across a hemisphere bounded by $\theta = \pi/2$ and located on the MKN black hole horizon. A fixed value of the magnetization parameter $\beta = 0.05$ has been selected. Projected contours are also shown for improved clarity of the plot. The surface is restricted by the condition for the emergence of the event horizon, $a^2 + e^2 \leq 1$. Four circles of $\sqrt{(a^2 + e^2)} = 0.25, 0.5, 0.75$, and 1.0 are shown to guide the eye. The yellow band on the surface, denoted by "Z", indicates where the total electric charge is zero. Note: Unlike the case of a weakly magnetized black hole, the moment of vanishing charge does *not* coincide with zero of the charge parameter, $e = 0$. On the other hand, $Q(a, e = 0)$ does not vanish and its graph is shown by a solid curve "Q". This is the feature of the exact MKN metric, where the two nulls do not generally coincide, as further detailed in [17] (this figure has been reproduced with permission from *Physica Scripta* article ref. [18]).

3. Weak Magnetic Field and Particle Acceleration

For the strong influence of the external magnetic field on the spacetime structure of the black hole, its intensity has to be enormously high, comparable with

$$B_{\text{GR}} = 10^{18} \, \frac{10 M_\odot}{M} \, [\text{G}]. \tag{13}$$

Realistic magnetic fields in astrophysical situations are strongly under this limit, even in the case of fields near magnetars, reaching $B \sim 10^{15}$ gauss. Therefore, for the astrophysical processes, we can usually put the magnetic spacetime factor $\Lambda = 1$ and the electric charge $e = 0$, using the canonical, asymptotically flat Kerr metric. As for the electromagnetic term, an asymptotically uniform magnetic field, orthogonal to the spacetime equatorial plane, can then be determined by the electromagnetic 4-vector potential taking the form

$$A_t = \frac{B}{2}(g_{t\phi} + 2a g_{tt}) - \frac{Q}{2} g_{tt} - \frac{Q}{2}, \quad A_\phi = \frac{B}{2}(g_{\phi\phi} + 2a g_{t\phi}) - \frac{Q}{2} g_{t\phi}, \tag{14}$$

where the induced electric charge of the black hole Q is also introduced. For non-charged black holes there is $Q = 0$, and the maximal induced black hole charge generated by the black hole rotation takes the Wald value $Q_W = 2aB$ (or $Q_W = 2aBM$ if we keep the mass term)—see [10]; the influence of the induced so-called Wald charge on the spacetime structure could be also abandoned [26,27]. For black holes with the maximal Wald charge we arrive at the electromagnetic potential

$$A_t = \frac{B}{2} g_{t\phi} - \frac{Q_W}{2}, \quad A_\phi = \frac{B}{2} g_{\phi\phi}. \tag{15}$$

It is crucial that even in this case the A_t component remains non-zero and can lead to a very strong acceleration mechanism for sufficiently massive black holes and strong magnetic fields [28]. The significant role of the electromagnetic fields in processes near a black hole horizon was for the first time presented in a series of works of Ruffini and his collaborators in [29]. It could be well demonstrated for the charged test particle motion in the case of ionized Keplerian disks [28].

The motion of an electrically charged test particle with charge q and mass m is determined by the Lorentz equation

$$m \frac{Du^\mu}{D\tau} = q F^\mu_\nu u^\nu, \tag{16}$$

where τ is the particle proper time, and F^μ_ν is the Faraday tensor of the electromagnetic field. For the Kerr–Newman black holes, the Lorentz equations can be separated and given in terms of first integrals, governing thus fully regular test particle motion [1,30,31], whereas for magnetized Kerr black holes, the separability is impossible implying a generally chaotic character of the motion [28,32–34].

Nevertheless, due to the symmetries of the magnetized Kerr black holes with the uniform magnetic field lines orthogonal to the equatorial plane of spacetime, we can introduce Hamiltonian in the form

$$H = \tfrac{1}{2} g^{\alpha\beta} (\pi_\alpha - q A_\alpha)(\pi_\beta - q A_\beta) + \tfrac{1}{2} m^2, \tag{17}$$

where the canonical four-momentum $\pi^\mu = p^\mu + q A^\mu$ is related to the kinematic four-momentum $p^\mu = m u^\mu$ and the influence of the electromagnetic field reflected by $q A^\mu$. The motion is then governed by the Hamilton equations

$$\frac{dx^\mu}{d\zeta} \equiv p^\mu = \frac{\partial H}{\partial \pi_\mu}, \quad \frac{d\pi_\mu}{d\zeta} = -\frac{\partial H}{\partial x^\mu}; \tag{18}$$

the affine parameter is related to the particle proper time as $\zeta = \tau/m$.

Due to the background symmetries, we can introduce two constants of the motion: energy E and angular momentum L as conserved components of the canonical momentum read

$$-E = \pi_t = g_{tt} p^t + g_{t\phi} p^\phi + q A_t, \tag{19}$$

$$L = \pi_\phi = g_{\phi\phi} p^\phi + g_{\phi t} p^t + q A_\phi. \tag{20}$$

Introducing the specific energy $\mathcal{E} = E/m$, the specific axial angular momentum $\mathcal{L} = L/m$, and the magnetic interaction parameter $\mathcal{B} = qB/2m$, we obtain Hamiltonian with two degrees of freedom, and the four-dimensional phase space $\{r, \theta; p_r, p_\theta\}$ in the form

$$H = \tfrac{1}{2} g^{rr} p_r^2 + \tfrac{1}{2} g^{\theta\theta} p_\theta^2 + \widetilde{H}_P(r, \theta), \tag{21}$$

enabling the introduction of the effective potential of the radial and latitudinal motion. The energy condition relates the specific energy to the effective potential as

$$\mathcal{E} = V_{\text{eff}}(r, \theta) \tag{22}$$

where

$$V_{\text{eff}}(r, \theta) = \frac{-\beta + \sqrt{\beta^2 - 4\alpha\gamma}}{2\alpha}, \tag{23}$$

with

$$\beta = 2[g^{t\phi}(\mathcal{L} - \tilde{q}A_\phi) - g^{tt}\tilde{q}A_t], \qquad \alpha = -g^{tt}, \tag{24}$$

and

$$\gamma = -g^{\phi\phi}(\mathcal{L} - \tilde{q}A_\phi)^2 - g^{tt}\tilde{q}^2 A_t^2 + 2g^{t\phi}\tilde{q}A_t(\mathcal{L} - \tilde{q}A_\phi) - 1. \tag{25}$$

The effective potential defined here is properly chosen for the region above the outer horizon of the black hole, governing the regions allowed for the motion of a charged particle with a fixed value of the axial angular momentum.

Study of the motion of charged particles applied to the case of ionized Keplerian disks (see [28] for a review) demonstrates that the fate of the ionized disks depends on the magnetic interaction parameter. In the so-called gravitational regime when gravity is suppressing the role of the electromagnetic field ($\mathcal{B} \ll 1$), the motion of the particles of the ionized Keplerian disks can be considered as being in quasi-circular harmonic epicyclic motion of regular character, enabling explanation of high-frequency quasi-periodic oscillations of X-rays observed in microquasars and some active galactic nuclei [35]. In the so-called gravity-magnetic regime when the role of both fields is comparable ($\mathcal{B} \sim 1$), the motion is fully chaotic, leading generally to toroidal configurations. In the so-called magnetic regime ($\mathcal{B} \gg 1$), the role of the magnetic field is decisive, and the motion could have finally a regular character governed by the Larmor precession frequency.

In the case of $\mathcal{B} > 1$ a special effect of chaotic scattering can be relevant [36,37] when the ionized particle can be accelerated along the magnetic field lines after a period of chaotic motion that decreases with increasing magnetic parameter [38]. In such situations, the magnetic Penrose process could be realized with extremely high efficiency. The tentative magnetic Penrose process (MPP; see [39]) is a local decay process; its energy balance is governed by the local value of the electromagnetic field (potential)—for this reason, the simple approximation of asymptotically uniform magnetic field aligned with the rotations axis can be well applied [28].

Let us consider the splitting of the 1st particle with energy E_1 (electrically neutral or positively charged with charge q_1) onto two charged particles, the 2nd one having a positive charge q_2 and the 3rd one having a negative charge q_3. If one of the particles (say the 3rd one) has a negative canonical energy $E_3 < 0$, then the second one should have the canonical energy $E_2 > E_1$ due to an extraction of the black hole energy because of the capture of the 3rd particle. The process of the split of the 1st particle into the 2nd and 3rd ones is governed by the conservation laws [39].

The efficiency of the MPP is defined by relating the gained and input energies

$$\eta = \frac{E_2 - E_1}{E_1} = \frac{-E_3}{E_1}, \tag{26}$$

implying the relation [40]

$$\eta_{\text{MPP}} = \chi - 1 + \frac{\chi q_1 A_t - q_2 A_t}{E_1}. \tag{27}$$

The MPP demonstrates three substantially different efficiency regimes. The low-efficiency regime corresponds to the original Penrose process involving only electrically neutral particles (or vanishing electromagnetic field) with efficiency [41]

$$\eta_{\text{PP(max)}} = \frac{\sqrt{2} - 1}{2} \sim 0.207. \tag{28}$$

The moderate regime of the MPP corresponds to the situation when the electromagnetic forces are dominant, and the particles are charged, i.e., the condition $|\frac{q}{m}A_t| \gg |u_t| = |p_t|/m$ is satisfied, with efficiency approximately determined as

$$\eta_{\text{MPP}}^{mod} \sim \frac{q_2}{q_1} - 1, \tag{29}$$

operating while $q_2 > q_1$. In this case, the gravitationally induced electric field of the black hole is neutralized and the moderate regime of the MPP is close to the Blandford–Znajek process [42]; both processes are driven by the quadrupole electric field generated due to twisting the magnetic field lines because of the spacetime frame dragging, and restricted by global neutrality of the plasma surrounding the black hole [3,43]. The extremely efficient regime corresponds to the ionization of neutral matter and its efficiency is dominated by the term

$$\eta_{\mathrm{MPP}}^{\mathrm{extr}} \sim \frac{q_2}{m_1} A_t. \tag{30}$$

In the extreme regime of the MPP, an enormous increase in the efficiency is possible, giving enormous energy to escaping particles. The efficiency can be as large as $\eta_{\mathrm{MPP}}^{\mathrm{extr}} \sim 10^{10}$ if the magnetic field is sufficiently large and the rotating black hole is supermassive [40] charging.

Let us note that the mechanism of charging of a boosted black hole in translatory motion has been revisited very recently, [44,45]; it has attracted renewed widespread attention because of its tentative relevance for late stages of black hole—neutron star inspirals and their subsequent mergers. In this context, there is an interesting parallel between the *effects of rotation vs. boost*. Along a different line of research, Okamoto and Song [46] argue that the electromagnetic self-extraction of energy will be possible only via the frame-dragged rotating magnetosphere. It will be interesting to see if the above-discussed ideas of *magnetic Penrose process*, where the energy extraction is explored from another view angle, will be confirmed with a more accurate and complete description in the future. It seems to be very exciting that the present-day understanding is still incomplete and even controversial as the adopted approximations are tentative and await further verification or disproval [47].

4. Conclusions

The MPP enables acceleration of protons and light ions up to the energy $E \sim 10^{22}$ eV, corresponding to the highest-energy ultra-high energy cosmic rays (UHECR) observed on the Earth, that can occur around supermassive black holes in the active galactic nuclei similar to those in the M87 large elliptical galaxy [40]. For accelerated electrons, the energy could be even higher, but contrary to the case of protons and ions, where the back-reaction related to the synchrotron radiation of the accelerated particles is negligible, for electrons the back-reaction is extremely strong, decelerating substantially this kind of light particles—they thus cannot be observed as UHECR [39].

Our scenario is complementary to highly dynamical situations discussed in a series of articles by Ruffini et al. [48], who explore the early, prompt phase of gamma-ray burst sources within a scenario of a baryonic shell interacting with an inhomogeneous medium (see also further references in [49–53]). Although we do not consider temporal effects on the black hole's gravitational field, we do take into account the role of the magnetic field in shaping the stationary background. It turns out that for astrophysically realistic models, time dependence may be crucial. On the other hand, the impact that super-strong magnetic fields may have on the spacetime curvature is relevant with respect to our understanding of exact solutions of Einstein–Maxwell fields; this can be best revealed by employing simplified equilibrium models such as the one discussed in our research note.

As a final remark, let us note that the similarity between the problem of a rotating magnetized body treated in the framework of classical electrodynamics and the corresponding black-hole electrodynamics has been widely explored in the literature (e.g., [54,55], and numerous subsequent papers). The black hole problem seems to be more complex because we have to consider the effects of general relativity; however, the adopted spacetime represents an electro-vacuum solution and it is thus idealized with a small number of free parameters. Intricate relations and numerical analysis are needed in order to determine material properties if plasma is present.

Author Contributions: Conceptualization, V.K.; Methodology, Z.S.; Writing—original draft, V.K.; Writing—review and editing, Z.S. All authors have read and agreed to the published version of the manuscript.

Funding: The authors thank the anonymous referees for their comments and suggestions that helped us to improve the paper and clarify several arguments. The authors acknowledge the institutional support from the Astronomical Institute in Prague and the Institute of Physics in Opava. V.K. acknowledges continued support from the Research Infrastructure CTA-CZ (LM2023047) funded by the Czech Ministry of Education, Youth and Sports, and the EXPRO (21-06825X) project of the Czech Science Foundation.

Data Availability Statement: No new data were created or analyzed in this study. Data sharing is not applicable to this article.

Acknowledgments: The authors are grateful to the organizers of the *Conference in Celebration of Remo Ruffini's 80th birthday*, held in the ICRANet seat at Villa Ratti, Nice (France) and online 16–18 May 2022, where the contribution was presented. It has been a great pleasure to meet and discuss with Ruffini during various occasions over the years in Rome, Pescara, and elsewhere, in particular at the *Marcel Grossmann Meetings on General Relativity*. More recently, RR and several colleagues of ICRANet have participated in the Relativistic Astrophysics Group Workshops *On Black Holes and Neutron Stars* (RAGtime) that are held annually at the Institute of Physics of the Silesian University in Opava, and at the *31st Texas Symposium on Relativistic Astrophysics* in Prague (Czech Republic).

Conflicts of Interest: The authors declare no conflict of interest.

References

1. Chandrasekhar, S. *The Mathematical Theory of Black Holes*; Oxford University Press: Oxford, UK, 1983.
2. DeWitt, C.; DeWitt, B.S. *Black Holes. Lectures Delivered at the Summer School of Theoretical Physics of the University of Grenoble at Les Houches*; Gordon and Breach: New York, NY, USA, 1973.
3. Punsly, B. *Black Hole Gravito-Hydromagnetics*; Springer: Berlin/Heidelberg, Germany, 2008.
4. Ruffini, R.; Wilson, J.R. Relativistic magnetohydrodynamical effects of plasma accreting into a black hole. *Phys. Rev. D* **1975**, *12*, 2959. [CrossRef]
5. Newman, E.T.; Couch, E.; Chinnapared, K.; Exton, A.; Prakash, A.; Torrence, R. Metric of a rotating, charged mass. *J. Math. Phys.* **1965**, *6*, 918. [CrossRef]
6. Baez, N.B.; García Díaz, A.G. The most general magnetized Kerr-Newman metric. *J. Math. Phys.* **1986**, *27*, 562. [CrossRef]
7. Ernst, F.J.; Wild, W.J. Kerr black holes in a magnetic universe. *J. Math. Phys.* **1976**, *12*, 1845. [CrossRef]
8. Kramer, D.; Stephani, H.; MacCallum, M., Herlt, E. *Exact Solutions of the Einstein's Field Equations*; Deutscher Verlag der Wissenschaften: Berlin, Germany, 1980.
9. Romero, G.E.; Vila, G.S. *Introduction to Black Hole Astrophysics*; Lecture Notes in Physics; Springer: Berlin/Heidelberg, Germany, 2014; Volume 876.
10. Wald, R.M. *General Relativity*; University of Chicago Press: Chicago, IL, USA, 1984.
11. Kerr, R.P. Gravitational field of a spinning mass as an example of algebraically special metrics. *Phys. Rev. Lett.* **1963**, *11*, 237. [CrossRef]
12. Gal'tsov, D.V. *Particles and Fields around Black Holes*; Moscow University Press: Moscow, Russia, 1986.
13. Christodoulou, D.; Ruffini, R. On the electrodynamics of collapsed objects. In *Black Holes*; DeWitt, C., DeWitt, B.S., Eds.; Gordon and Breach Science Publishers: New York, NY, USA, 1973; p. R151
14. Kinnersley, W. Generation of stationary Einstein-Maxwell fields. *J. Math. Phys.* **1973**, *14*, 651. [CrossRef]
15. García Díaz, A. Magnetic generalization of the Kerr-Newman metric. *J. Math. Phys.* **1985**, *26*, 155. [CrossRef]
16. Hiscock, W.A. On black holes in magnetic universes. *J. Math. Phys.* **1981**, *22*, 1828. [CrossRef]
17. Karas, V.; Vokrouhlický, D. On interpretation of the magnetized Kerr-Newman black hole. *J. Math. Phys.* **1990**, *32*, 714. [CrossRef]
18. Karas, V.; Budínová, Z. Magnetic fluxes across black holes in a strong magnetic field regime. *Phys. Scr.* **2000**, *61*, 253. [CrossRef]
19. Karas, V. Magnetic fluxes across black holes. Exact models. *Bull. Astron. Inst. Czechoslov.* **1988**, *39*, 30.
20. Ruffini, R. On gravitationally collapsed objects. In *Relativity, Quanta and Cosmology in the Development of the Scientific Thought of Albert Einstein*; Pantaleo, M., de Finis, F., Eds.; Johnson Reprint Corp.: New York, NY, USA, 1979; pp. 599–658.
21. Damour, T.; Hanni, R.S.; Ruffini, R.; Wilson, J.R. Regions of magnetic support of a plasma around a black hole. *Phys. Rev. D* **1978**, *17*, 1518. [CrossRef]
22. Hanni, R.S.; Valdarnini, R. Magnetic support near a charged rotating black hole. *Phys. Lett. A* **1979**, *70*, 92. [CrossRef]
23. Lyutikov, M. Schwarzschild black holes as unipolar inductors: Expected electromagnetic power of a merger. *Phys. Rev. D* **2011**, *83*, 064001. [CrossRef]
24. Morozova, V.S.; Rezzolla, L.; Ahmedov, B.J. Nonsingular electrodynamics of a rotating black hole moving in an asymptotically uniform magnetic test field. *Phys. Rev. D* **2014**, *89*, 104030. [CrossRef]

25. Karas, V.; Vokrouhlický, D. Dynamics of charged particles near a black hole in a magnetic field. *J. Phys. I France* **1991**, *1*, 1005. [CrossRef]
26. Bičák, J.; Dvořák, L. Stationary electromagnetic fields around black holes. III. *Phys. Rev. D* **1980**, *22*, 2933. [CrossRef]
27. King, A.R.; Lasota, J.P.; Kundt, W. Black holes and magnetic fields. *Phys. Rev. D* **1975**, *12*, 3037. [CrossRef]
28. Stuchlík, Z.; Kološ, M.; Kovář, J.; Slaný, P.; Tursunov, A. Influence of cosmic repulsion and magnetic fields on accretion disks rotating around Kerr black holes. *Universe* **2020**, *6*, 26. [CrossRef]
29. Ruffini, R. On the energetics of black holes. In *Black Holes (Les Astres Occlus), Lectures Delivered at the Summer School of Theoretical Physics of the University of Grenoble at Les Houches*; DeWitt, C., DeWitt, B.S., Eds.; Gordon and Breach: New York, NY, USA, 1973; pp. 451–546.
30. Carter, B. Black hole equilibrium states. In *Black Holes (Les Astres Occlus), Lectures Delivered at the Summer School of Theoretical Physics of the University of Grenoble at Les Houches*; DeWitt, C., DeWitt, B.S., Eds.; Gordon and Breach: New York, NY, USA, 1973; pp. 57–214.
31. Wald, R.M. On uniqueness of the Kerr-Newman black holes. *J. Math. Phys.* **1972**, *13*, 490. [CrossRef]
32. Lukes-Gerakopoulos, G. Adjusting chaotic indicators to curved spacetimes. *Phys. Rev. D* **2014**, *89*, 043002. [CrossRef]
33. Pánis, R.; Kološ, M.; Stuchlík, Z. Determination of chaotic behaviour in time series generated by charged particle motion around magnetized Schwarzschild black holes. *Eur. Phys. J. C* **2019**, *79*, 479. [CrossRef]
34. Karas, V.; Vokrouhlický, D. Chaotic motion of test particles in the Ernst space-time. *Gen. Relativ. Gravit.* **1992**, *24*, 729–743. [CrossRef]
35. Stuchlík, Z.; Kološ, M.; Tursunov, A. Large-scale magnetic fields enabling fitting of the high-frequency QPOs observed around supermassive black holes. *Publ. Astron. Soc. Jpn.* **2022**, *74*, 1220. [CrossRef]
36. Kopáček, O.; Karas, V.; Kovář, J.; Stuchlík, Z. Transition from regular to chaotic circulation in magnetized coronae near compact objects. *Astrophys. J.* **2010**, *722*, 1240. [CrossRef]
37. Stuchlík, Z.; Kološ, M. Acceleration of the charged particles due to chaotic scattering in the combined black hole gravitational field and asymptotically uniform magnetic field. *Eur. Phys. J. C* **2016**, *76*, 32. [CrossRef]
38. Karas, V.; Kopáček, O. Near horizon structure of escape zones of electrically charged particles around weakly magnetized rotating black hole: Case of oblique magnetosphere. *Astron. Nachrichten* **2021**, *342*, 357. [CrossRef]
39. Stuchlík, Z.; Kološ, M.; Tursunov, A. Penrose process: its variants and astrophysical applications. *Universe* **2021**, *7*, 416. [CrossRef]
40. Tursunov, A.; Stuchlík, Z.; Kološ, M. Supermassive black holes as possible sources of ultrahigh-energy cosmic rays. *Astrophys. J.* **2020**, *895*, 14. [CrossRef]
41. Penrose, R. Gravitational collapse: The role of General Relativity. *Riv. Nuovo C.* **1969**, *1*, 252.
42. Blandford, R.D.; Znajek, R.L. Electromagnetic extraction of energy from Kerr black holes. *Mon. Not. R. Astron. Soc.* **1977**, *179*, 433. [CrossRef]
43. Dadhich, N.; Tursunov, A.; Ahmedov, B.; Stuchlík, Z. The distinguishing signature of magnetic Penrose process. *Mon. Not. R. Astron. Soc. Lett.* **2018**, *478*, L89. [CrossRef]
44. Dai, Z.G. Inspiral of a spinning black hole-magnetized neutron star binary: Increasing charge and electromagnetic emission. *Astrophys. J. Lett.* **2019**, *873*, L13. [CrossRef]
45. Adari, P.; Berens, R.; Levin, J. Charging up boosted black holes. *Phys. Rev. D* **2023**, *107*, 044055. [CrossRef]
46. Okamoto, I.; Song, Y. Energy self-extraction of a Kerr black hole through its frame-dragged force-free magnetosphere. *arXiv* **2023**, arXiv:1904.11978.
47. Santos, J.; Cardoso, V.; Natário, J. Electromagnetic radiation reaction and energy extraction from black holes: The tail term cannot be ignored. *Phys. Rev. D* **2023**, *107*, 064046. [CrossRef]
48. Ruffini, R.; Bernardini, M.G.; Bianco, C.L. On gamma-ray bursts. In Proceedings of the Eleventh Marcel Grossmann Meeting on Recent Developments in Theoretical and Experimental General Relativity, Gravitation and Relativistic Field Theories, Berlin, Germany, 23–29 July 2006; Kleinert, H., Jantzen, R.T., Ruffini, R., Eds.; World Scientific Publishing Co.: Singapore, 2008.
49. Fryer, C.L.; Rueda, J.A.; Ruffini, R. Hypercritical accretion, induced gravitational collapse, and binary-driven hypernovae. *Astrophys. J. Lett.* **2014**, *793*, L36. [CrossRef]
50. Komissarov, S.S. Electrically charged black holes and the Blandford-Znajek mechanism. *Mon. Not. R. Astron. Soc.* **2022**, *512*, 2798. [CrossRef]
51. Metzger, B.D. Luminous fast blue optical transients and type Ibn/Icn SNe from Wolf-Rayet/Black hole mergers. *Astrophys. J.* **2022**, *932*, 84. [CrossRef]
52. Rueda, J.A.; Ruffini, R. On the induced gravitational collapse of a neutron star to a black hole by a type Ib/c supernova. *Astrophys. J. Lett.* **2012**, *758*, L7. [CrossRef]
53. Ruffini, R.; Wang, Y.; Aimuratov, Y.; de Almeida, U.B.; Becerra, L.; Bianco, C.L.; Xue, S.S. Early X-ray flares in GRBs. *Astrophys. J.* **2018**, *852*, 53. [CrossRef]

54. Ruffini, R.; Treves, A. On a magnetized rotating sphere. *Astrophys. Lett.* **1973**, *13*, 109.
55. Tursunov, A.; Zajaček, M.; Eckart, A.; Kološ, M.; Britzen, S.; Stuchlík, Z.; Karas, V. Effect of electromagnetic interaction on Galactic center flare components. *Astrophys. J.* **2020**, *897*, 99. [CrossRef]

universe

Review

Eclipses: A Brief History of Celestial Mechanics, Astrometry and Astrophysics

Costantino Sigismondi [1,2,3,*,†] and Paolo De Vincenzi [4,*,†]

1   ICRANet International Center for Relativistic Astrophysics Network, Piazza della Repubblica 10, 65122 Pescara, Italy
2   APRA—Ateneo Pontificio Regina Apostolorum Via degli Aldobrandeschi 190, 00163 Roma, Italy
3   ITIS G. Ferraris, Via Fonteiana 111, 00152 Roma, Italy
4   Dipartimento di Fisica, Sapienza Università di Roma, Piazzale A. Moro 2, 00185 Roma, Italy
*   Correspondence: sigismondi@icra.it (C.S.); paolo.devincenzi@uniroma1.it (P.D.V.)
†   These authors contributed equally to this work.

Abstract: Solar and lunar eclipses are indeed the first astronomical phenomena which have been recorded since very early antiquity. Their periodicities gave birth to the first luni-solar calendars based on the Methonic cycle since the sixth century before Christ. The Saros cycle of 18.03 years is due to the Chaldean astronomical observations. Their eclipses' observations reported by Ptolemy in the Almagest (Alexandria of Egypt, about 150 a.C.) enabled modern astronomers to recognize the irregular rotation rate of the Earth. The Earth's rotation is some hours in delay after the last three millenia if we use the present rotation to simulate the 721 b.C. total eclipse in Babylon. This is one of the most important issues in modern celestial mechanics, along with the Earth's axis nutation of 18 yr (discovered in 1737), precession of 25.7 Kyr (discovered by Ipparchus around 150 b.C.) and obliquity of 42 Kyr motions (discovered by Arabic astronomers and assessed from the Middle Ages to the modern era, IX to XVIII centuries). Newtonian and Einstenian gravitational theories explain fully these tiny motions, along with the Lense–Thirring gravitodynamic effect, which required great experimental accuracy. The most accurate lunar and solar theories, or their motion in analytical or numerical form, allow us to predict—along with the lunar limb profile recovered by a Japanese lunar orbiter—the appearance of total, annular solar eclipses or lunar occultations for a given place on Earth. The observation of these events, with precise timing, may permit us to verify the sphericity of the solar profile and its variability. The variation of the solar diameter on a global scale was claimed firstly by Angelo Secchi in the 1860s and more recently by Jack Eddy in 1978. In both cases, long and accurate observational campaigns started in Rome (1877–1937) and Greenwich Observatories, as well as at Yale University and the NASA and US Naval Observatory (1979–2011) with eclipses and balloon-borne heliometric observations. The IOTA/ES and US sections as well as the ICRA continued the eclipse campaigns. The global variations of the solar diameter over a decadal timescale, and at the millarcsecond level, may reflect some variation in solar energy output, which may explain some past climatic variations (such as the Allerød and Dryas periods in Pleistocene), involving the outer layers of the Sun. "An eclipse never comes alone"; in the eclipse season, lasting about one month, we can have also lunar eclipses. Including the penumbral lunar eclipses, the probability of occurrence is equi-distributed amongst lunar and solar eclipses, but while the lunar eclipses are visible for a whole hemisphere at once, the solar eclipses are not. The color of the umbral shadow on the Moon was known since antiquity, and Galileo (1632, Dialogo sopra i Massimi Sistemi del Mondo) shows clearly these phenomena from copper color to a totally dark, eclipsed full Moon. Three centuries later, André Danjon was able to correlate that umbral color with the 11-year cycle of solar activity. The forthcoming American total solar eclipse of 8 April 2024 will be probably the eclipse with the largest mediatic impact of the history; we wish that also the scientific impulse toward solar physics and astronomy will be relevant, and the measure of the solar diameter with Baily's beads is indeed one of the topics significantly related to the Sun–Earth connections.

Keywords: eclipses, lunar and solar; occultations; Baily's beads; solar diameter

Citation: Sigismondi, C.; De Vincenzi, P. Eclipses: A Brief History of Celestial Mechanics, Astrometry and Astrophysics. *Universe* **2024**, *10*, 90. https://doi.org/10.3390/universe10020090

Received: 25 November 2023
Revised: 17 January 2024
Accepted: 31 January 2024
Published: 13 February 2024

1. Introduction

The Sun is the closest star to our planet. The accurate knowledge of its physics is also knowledge of our living environment because of the strong energetic relationship with our climate. Most of the new physics of the XX century is involved in the study of the Sun: General Relativity, Nuclear Physics and Neutrino phase mixing physics. These three domains of modern physics considered the Sun a privileged laboratory. The secular variations of the solar diameter as well as the unexplicable great dimming of solar activity, known as Maunder minimun of 1645–1715 or the great maxima and minima identified in the past millenia [1], do not find a suitable predicting model. In this paper, we briefly review the history of the eclipses' science, both solar and lunar, from antiquity to the present days, focusing on the solar diameter measurements at the lunar shadow's limits through Baily's beads. The variability of the solar diameter is sustained by several accurate measures in the last 130 years as well as from historical eclipses since 1567. In this general context, as celestial mechanics, we consider mainly the geometrical models yielding the predictions of the solar and lunar positions, especially the ones (Methon and Saros "circular motions") which predict eclipses within a given range of decades or centuries.

2. Historical Solar Eclipses and Implications in Celestial Mechanics

2.1. Chaldean Astronomy

For millenia, mankind turned his gaze to the sky to admire and study its wonders. Of all the phenomena, solar and lunar eclipses are those that have influenced the mythology, religion and science of early civilizations the most. Chaldeans identified the periodic recurrence of lunar and solar eclipses, calculating a cycle of 18 years, 10/11 days, 8 h and 42 min, the Saros. Through prolonged and careful observations, the Chaldeans were not only able to predict the following eclipses; they also noted that the same eclipse occurred in the same place after 3 Saros, an Exeligmos cycle of 54 years and 34 days, as also described by Ptolemy in his writings [2–4]. It is safe to assume that the geometrical nature of the eclipses was already clear at that time. In general, the occultation of a body (the Sun, a star or a planet) by the Moon or an asteroid provides information on the shape of the darkening body itself and the relative distance between it and the darkened one (Figure 1). Also, the occultation of the Sun by the Earth produces, on the Moon, a shadow, which was called "defectus", since the Moon is not occulted to our sight.

Figure 1. Venus' occultation of 9 November 2023 at 10:09:48 UT observed in Rome, via Fonteiana 111.

2.2. Occultation'S Astronomy

Figure 1 helps explain how occultations are related to celestial mechanics as it displays Venus' occultation on 9 November 2023. The instant of the disappearance of Venus has been determined within ± 0.01 s of accuracy, which allows for locating the Moon with ± 10 m of space accuracy and 3 millarcsec of angular accuracy [5].

Asteroidal occultations are currently used to assess the asteroidal orbits with great accuracy and, in the case of some stars with large angular diameters like Betelgeuse (60 mas) and Regulus (1.7 mas), also their limb-darkening functions [6]. The eclipses of the Galilean satellite io by Jupiter offered the possibility to measure the speed of light by Roemer on his and Cassini's data (1676) [7].

2.3. Eclipses' Astronomy and Earth's Rotation Rate

In the Sun–Earth–Moon system, two types of eclipses are recognized: lunar and solar. The first occurs when the Moon, whose orbital plane is inclined by 5.9° compared to the ecliptic, passes through one of the intersection nodes of the orbit in opposition to the Sun while being hit by the cone of shadow projected by our planet. These eclipses are visible from any point on the Earth's hemisphere where the Moon is above the horizon. Solar eclipses, on the other hand, occur when the passage through one of the nodes occurs with the Moon in conjunction. The Moon projects its shadow on our planet, and due to the relative size and distances between the Sun, the Earth and itself, the shadow cone can vary in size and duration, giving information on the position of the three bodies in space. The most studied type of eclipse is the total solar eclipse, which is visible when the Earth–Moon distance is such that the angular diameter of our satellite is slightly greater than that of the Sun. This suggestive event has allowed also the first studies on the solar corona. The shadow cone generated during these phenomena covers a narrow band of the Earth's surface, less than few hundreds of kilometers wide, where it is possible to observe a sudden and great variation in brightness of the sky (from 10 magnitudes to 10,000 times in intensity). Ptolemy reported in the Almagest (150 a.C.) the total solar eclipse recorded by the Chaldeans in 721 b.C. [8,9]. The occurrence of this phenomenon in Babylon tells us that the current Earth's rotation rate changed during the last 27 centuries due to the ongoing post-glacial isostatic rebound of the continents [10–12].

2.4. Earth's Axis Millennial Motions

The luni-solar precession (or "of the equinoxes") is responsible for the double-conical movement of the Earth's axis. With a period of approximately 25,700 years, the combined gravitational action of the Sun and Moon on the equatorial bulge tends to align the planet's rotation axis along the direction perpendicular to the ecliptic plane. This gravitational pull is opposed by the rotational movement of the Earth, which keeps the angular momentum fixed. The resulting effect is a shift of the equinoxes by 20 arcminutes westwards [13]. Together with the precession of the equinoxes, and again due to the effect of the tidal forces deriving from the action of the Sun and the Moon, a second motion of our planet is observed: the nutation, which was discovered in 1737 by James Bradley. Nutation is characterized by a subtle wobble in the Earth's rotation axis, which follows a cycle of approximately 18.6 years. The maximum amplitude of nutation in ecliptic latitude is 9 arcseconds. The Earth's obliquity also variates in 42 Ky of $\pm 2°$. Since the dawn of observative astronomy, the Earth's axis changed from about 24° to the present 23.42°.

2.5. Solar Astrometry: The Eclipse of Clavius

Another example of deducing changes in the properties of a celestial body using an eclipse is represented by the observation of the annular–total solar eclipse in Rome in 1567 AD by the Jesuit mathematician Clavius [14]. He personally observed the 1567 eclipse (9 May) and the eclipse of Coimbra in 1560 (21 August), and he reported, for the one in 1567, the presence of a clear disk of light around the Moon. Clavius deduced that the angular diameter of the Sun was greater than that of our satellite, which went against Ptolemy and medieval Arab astronomers. The nature of the ring of light observed by Christopher Clavius has been investigated by Kepler and by subsequent studies [15,16]. The occurrence of an annular eclipse in 1567 in Rome, instead of total, as predicted by the ephemerides for that day in Rome using current parameters, remains still intriguing. The solar diameter should have been 0.2% larger than expected, and the 1567 eclipse's discussions gave birth to the studies on the secular variations of the diameter of the Sun [17]. It is therefore clear that not only do eclipses give important information on the celestial mechanics of the bodies involved, but they can also shed light on their properties and characteristics and how these influence our planet.

3. Lunar Eclipses from the Metonic Cycle to the Solar Corona Reflection

The celestial mechanics of the Moon are at the basis of civil calendars because, since ancient times, it has been necessary to identify a univocal criterion for organizing religious celebrations and planning social events. The Babylonians were the first to create a calendar based on the lunar motion [18]. However, they were unable to reconcile the discrepancies between the tropical year and the draconic year. The civil year has to correspond to the tropical one, which lasts 365.242 days, or the time between two consecutive Spring equinoxes. The draconic year is the time taken by the Sun to return to the same node of the lunar orbit. Since the line of intersection of the ecliptic plane with that of the lunar orbit moves retrogradely, the position of the nodes moves back annually by $19°33'$, and the Sun does not meet the same node again after a tropical year but rather approximately 19 days earlier, i.e., every 346.62 days. The first to create a synchronized calendar was the Greek astronomer Meton (432 b.C.), who, through the observation of 19 consecutive solar years (corresponding to approximately 235 lunar months and 6940 days) and starting from the Saros cycle, created a calendar so that the motion of the two celestial bodies would return in phase. Numerous calendars from the classical and medieval era are based on it and were later modified according to the uses and needs of each culture [19–21]. The discovery of the Metonic cycle was probably the necessary stimulus to introduce leap years to recover the decimal part of the solar year which is suppressed in the calculation of the civil year, which is made necessarily of an integer number of days. There are testimonies of reform attempts already in the Ptolemaic Egypt [21], but the introduction of a temporally synchronized system is due to Julius Caesar. He, just after having been in Alexandria, introduced the Julian calendar to the Roman Senate in 46 b.C., and it became the basis of the modern Western calendar.

3.1. The Metonic Cycle

The lunar calendar and the Metonic cycle were used by the Catholic Church for the Easter's Computus, the determination of the day of Easter, and it is still used after the Gregorian reformation of the calendar. Although in continental Europe the scientific debate had slowed down in the last centuries of the Roman Empire, the Church made numerous efforts to codify the calculation of the celebration of the resurrection. The principle-rule that fixes the date of Christian Easter was established following the Council of Nicea (325 AD) [22]: Easter falls on the Sunday following the first full moon of spring (at the time of the first calculations, the equinox fell on 21 March, which therefore became the reference date). Consequently, the Easter date is always included in the period from 22 March to 25 April. It is important to point out that among the first controversies in the Catholic Church, there are the ones on the Easter's algorithms based on the local traditions in the geographical area concerned [23–26].

3.2. Lunar Secular Motions

The lunar motion has more than 400 terms in the modern ephemerides, but for eclipses and solar astrometry, we recall the precession of the lunar axis and the variation in the eccentricity of the orbit of our satellite. The improvement of telescopes in the XVII century allowed a more accurate study of such motions. Among these there are librations: apparent movements of the Moon that allow an observer on Earth to see slightly different portions of the lunar surface each time. These variations are caused by the fact that the Moon rotates around its axis at a constant rate but revolves around the Earth at a variable rate, being in an elliptical orbit and moving faster when it is closer to the Earth and slower when it is further away from it. The final effect is that instead of half, only 41% of the lunar surface is always visible, another 41% is always hidden, and a further 18% oscillates between the visible and hidden portions of the surface bringing to 59% the total visible surface of the Moon over the course of an entire libration cycle [27]. The total oscillation effect is given by the contribution of the two previously introduced motions: the inclination of the lunar axis and the libration in latitude discovered by Galileo in 1632 and the variations in the eccentricity of the orbit and the libration in longitude discovered by Hevelius in 1648 [28,29]. The lunar profile during solar total eclipses is of course subjected to all these motions.

3.3. Lunar Colors during the Eclipses from Galileo to Danjon

Galileo Galilei in his writings summarized the discoveries made possible by the telescope, undermining the Ptolemaic certainties on the celestial order: the existence of the seas and the lunar craters and the satellites orbiting around Jupiter were powerful arguments in favor of the Copernican theory [30–33]. In the "Saggiatore" (1623), dedicated to the nature of the comets as real celestial bodies, Galileo also dealt with the colors assumed by the darkened part of the Moon [34,35], as can be seen in Figure 2. It was only in 1921 that the astronomer André-Louis Danjon, through an intensity scale of the eclipsed Moon brightness, was able to connect such phenomenon to the solar activity [36]. The color variations of the shadowed area of the Moon are certainly influenced by atmospheric conditions and by the light reflected from the Earth's surface, but they depend above all on the extension of the solar corona associated with the various phases of our star's activity cycle, as already understood by Angelo Secchi in the second half of the XIX century [37–41].

Figure 2. Partial lunar eclipse of 29 October 2023 at 20:36:37 UT, observed in Ostia.

4. The Sun–Earth Connection

Solar eclipses have been for a long time the only means available for studying the solar corona. This changed with the advent of Bernard Lyot's coronagraph, which is an instrument used to simulate eclipses that can also be employed on satellites dedicated to solar observation, such as the SkyLab and SOHO [42–44]. The study of the Sun from space has also allowed researchers to shed light on other phenomena known since ancient times: the polar auroras. The interaction of high-energy charged particles, components of the solar wind, with the Earth's ionosphere gives life, through radiative de-excitation mechanisms, to the characteristic colored and bright bands in the polar sky, where the shielding effect of the Earth's magnetic field is less effective. The first to relate solar activity to auroras was the astronomer R. C. Carrington, who observed a brief bright light in a group of unusually large sunspots on 31 August 1859 [45–47]. On that day, polar auroras were visible all across the planet, causing the generation of spurious currents in electric circuits, damaging them [48]. This event has been later associated with a powerful solar flare, exactly Earth-facing: it is known as the Carrington event and was also measured by Angelo Secchi's magnetometers in Rome [49]. Since then, studying and monitoring sunspots has played a particularly important role in predicting this type of particles storms, as it would allow us to take the right countermeasures to protect our, now very dense, telecommunications network [50]. Although the nature of sunspots is not yet fully understood, nowadays, we know that they are formed following a decrease in the energy arriving from the star's core to the surface due to a variation in the magnetic field caused by the differential rotation of the celestial body. In these areas of lower temperature (around 4000 K compared to the 6000 K of the surrounding photosphere) a self-sustaining mechanism is established, similar to that of hurricanes, due to the strong magnetic fields. The occurrence of sunspots is linked to the main eleven-year cycle of solar activity; their existence has been known since 800 b.C. [51] and has been recorded since then by various astronomers (Galileo included) [52,53]. They are therefore a good indicator of solar activity throughout the centuries, and their observation will be useful to better understand the internal mechanisms of our star and how, and if, they affect our planet. Cycles longer than the 11-years have also been identified (80–100 years, 800–1200 years, etc.), and they are also linked to the planetary periodicities [54].

5. Recent Eclipses: Preliminary Results on the Solar Radius

In the last few decades, the number of eclipse observers has constantly increased, along with the quality level of their observations. After the first missions to the shadows' limbs in the 1970s [55,56], the solar diameter has been monitored with this method several times; an example is in Figure 3. An Atlas of observed Baily's beads [57] was collected and used to recover the solar limb-darkening function [58] and its inflexion point, which unifies this approach with the solar limb definition used in the oblateness measures since Robert Dicke's in 1967 [59–61].

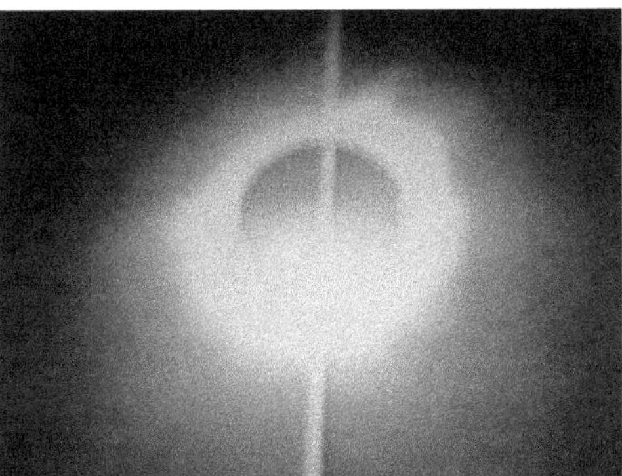

Figure 3. The long-lasting final Baily's bead at the shadow's limit in Egypt as observed by Zawyet al Mahtallah on 29 March 2006, visible with the full corona [62,63].

On the lunar shadow's limits, the Baily's beads last longer, making this method more accurate, even if it is subjected to the filter's cutoff [63]. For this reason, the International Occultation Timing Association (IOTA) adopted in 2010 a standard filter to fix also the wavelength at which the diameter is measured (520 nm), as in the Solar Disk Sextant SDS [64] balloon-borne mission.

5.1. Solar Diameter's Standard Value

In the last decade, the eclipses, lunar and solar showed a solar radius around 960.0″, which is larger than the 959.63″ IAU standard value [65] at 1 AU. The values of the solar diameter here discussed all refer to 1 AU, so any variation of the Earth–Sun distance along the year or the eccentricity's changes in the orbit does not affect these values; only intrinsic variations do.

5.2. Solar Diameter's Variations

The variation of the solar diameter was confirmed also with photometers' arrays displaced from French missions from 2012 to 2015 [66] in another experiment designed to avoid the cutoff problem. During 2023, two eclipse have occurred: a hybrid on 20 April [67] and an annular on 14 October. In both cases, the preliminary results confirm the solar radius to be in the 960.0″ range [68], with a technique also used in 2022 as can be seen in Figure 4 The statement about the need to change the IAU standard value of 959.63″ dates back to 1891, and it does not give enough importance to the real change of the solar diameter. The measures of 1891 were performed with excellent optics, and careful methods, and they were confirmed by many eclipses, lunar and solar' analyses and by SDS flights. The solar diameter was measured within ±0.02″ and increased from 959.63″ (1992) to 959.86″ (2011) [64]. The increase in the solar diameter was also detected with the Danjon-modified solar astrolabes (1975–2009) in France, Algery, Turkey, Spain and Brazil [69,70]. Another instrument, the reflecting heliometer, was developed in Rio de Janeiro to monitor the solar diameter especially during the coronal mass ejections, which are the major events in space weather [71]. Other measures in different wavelengths [72] confirmed the diameter's enlargement. The stellar standard model does not explain global solar diameter oscillations on yearly scales but rather only the helioseismic waves of 5 min [73].

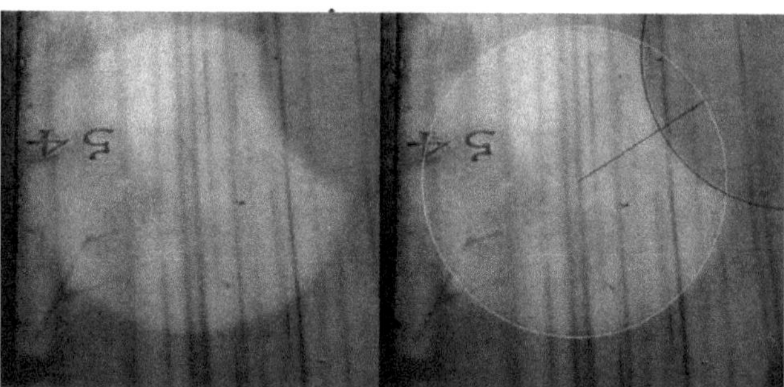

Figure 4. The solar partial eclipse of 25 October 2022 at 10:54:20 UT on the Clementine meridian line of 1702 in Rome [74]. This eclipse was a real rare partial one for all the World [75].

6. Conclusions and Perspectives

The solar activity was identified in the XIX century with the 11-year sunspots period and was later corrected to a 22-year cycle after Hale magnetic observations [76]. The corresponding coronal activity was first identified by Secchi [49] and subsequently confirmed by the observations during the eclipses, lunar and solar and with coronographs. Later, at the birth of helioseismology, global solar oscillations of 5 min were detected [77]. A gap of seven decades (1645–1715) in the solar activity was detected by Maunder [78], but the first idea to correlate sunspots activity to the solar diameter variations came out after 1978 with the analysis of Clavius' eclipse of 1567 observed in Rome. Since then, the total eclipses, lunar and solar' accounts have been exploited to measure the solar diameter thanks to the rapid luminosity variation occurring near the solar limb in the last arcseconds. Ancient data were obtained with the naked eye, while photo, video and electronic devices prevailed in the last few decades. General consensus in the present day (2023) establishes the solar radius at $960.0''$ not only from eclipse data, and the solar radius increased by about $0.4''$ over the last 130 years. Ancient and recent planetary transits have been used to assess this statement also with SOHO and SDO satellites [79,80]; however, there is a lack of consensus here. The instruments devoted to real-time solar diameter's measurement, the reflecting heliometer of Rio de Janeiro and the solar astrolabes in Nice (DORAYSOL) and Rio are currently off duty. There is no news from the other instrument of Nice/Calern Observatory: Picard-Sol [81]. The fascinating eclipses' missions, once possible as a national effort, are now operated also by an increasing number of valent amateur and professional astronomers, who are inspired in their actions by the unforgettable Jay Myron Pasachoff (1943–2022) and Serge Koutchmy (1940–2023) who were able to set new astrophysical experiments for each new eclipse since their first observational missions [82]. The quality of the new incoming data is guaranteeing the validity of the forthcoming research on the secular variability of the solar diameter and on the correlation of the diameter with other observables more closely related with the Earth's climate variability.

Author Contributions: Resources: C.S. and P.D.V. Writing—original draft preparation C.S. and P.D.V. Writing—review and editing C.S. and P.D.V. All authors have read and agreed to the published version of the manuscript.

Funding: This research received no external funding.

Data Availability Statement: The observations quoted in the text and represented in the figures are published in the youtube astronomical archive of Costantino Sigismondi, YouTube: https://www.youtube.com/channel/UCe18v3EZ8w2qmd8jW6mYV5w (accessed on 30 January 2024).

Conflicts of Interest: The authors declare no conflicts of interest.

References

1. Usoskin, I. A history of solar activity over millennia. *Living Rev. Sol. Phys.* **2023**, *20*, 2. [CrossRef]
2. Pinches, T.G.; Strassmaier, J.N. *Late Babylonian Astronomical and Related Texts*; Sachs, A.J., Ed.; Brown University Press: Providence, RI, USA, 1955.
3. Sachs, A.J.; Hunger, H. *Astronomical Diaries and Related Texts from Babylonia*; Verlag der Österreichischen Akademie der Wissenschaften: Vienna, Austria, 1988; Volume 1.
4. Huber, P.J.; De Meis, S.; Goldstein, B.R. Babylonian Eclipse Observations from 750 BC to 1 BC. In *Aestimatio: Sources and Studies in the History of Science*; IsIAO/Mimesis: Milano, Italy, 2004; pp. 122–125.
5. Sigismondi, C. Relativistic Corrections to Lunar Occultations. *J. Korean Phys. Soc.* **2010**, *56*, 1694–1699. [CrossRef]
6. Sigismondi, C.; Costa, C.; Noschese, A.; Guhl, K.; Bisconte, M. Signal-to-Noise Improvements for Observations of 2023 Betelgeuse's Occultation. *J. Occult. Astron.* **2024**, *14*, 27–31.
7. Tuinstra, F. Rømer and the finite speed of light. *Phys. Today* **2004**, *57*, 16–17. [CrossRef]
8. Heiberg, J.L. *Claudii Ptolemaei Opera Quae Exstant Omnia: 1 Syntaxis Mathematica*; Teubner: Leipzig, Germany, 1898.
9. Toomer, G.J. *Ptolemy's Almagest*; Princeton University Press: Princeton, NJ, USA, 1998.
10. Stephenson, F.R.; Morrison, L.V.; Smith, F.T. Long-term fluctuations in the Earth's rotation: 700 BC to AD 1990. *Philos. Trans. R. Soc. Lond. Ser. Phys. Eng. Sci.* **1995**, *351*, 165–202.
11. Stephenson, F.R. Babylonian Timings of Eclipse Contacts and the Study of the Earth's Past Rotation. *J. Astron. Hist. Herit.* **2006**, *9*, 145–150. [CrossRef]
12. Yoder, C.F.; Williams, J.G.; Dickey, J.O.; Schutz, B.E.; Eanes, R.J.; Tapley, B.D. Secular variation of Earth's gravitational harmonic J_2 coefficient from Lageos and nontidal acceleration of Earth rotation. *Nature* **1983**, *303*, 757–762. [CrossRef]
13. Hohenkerk, C.; Yallop, B.D.; Smith, C.A.; Sinclair, A.T. Celestial Reference Systems. In *Explanatory Supplement to the Astronomical Almanac*; Seidelmann, P.K., Ed.; University Science Books: Sausalito, CA, USA, 1992.
14. Sigismondi, C. Incontri celesti, vita del padre Clavio in cinque atti. *arXiv* **2011**, arXiv:1106.2517.
15. Sigismondi, C. Christopher Clavius astronomer and mathematician. *arXiv* **2012**, arXiv:1203.0476.
16. Eddy, J.; Boornazian, A.A.; Clavius, C.; Shapiro, I.I.; Morrison, L.V.; Sofia, S. Shrinking Sun. *Sky Telesc.* **1980**, *60*, 10.
17. Eddy, J.A. Climate and the Role of the Sun. *J. Interdiscip. Hist.* **1980**, *10*, 725–747. [CrossRef]
18. Neugebauer, O. *The Exact Sciences in Antiquity*; Courier Corporation: North Chelmsford, MA, USA, 1969; Volume 9.
19. Merrit, B.D. *The Athenian Year, Berkeley*; University of California: Los Angeles,CA, USA, 1961.
20. Mosshammer, A. *The Easter Computus and the Origins of the Christian Era*; OUP Oxford: Oxford, UK, 2008.
21. Stern, S. *Calendars in Antiquity (Empires, States and Societies)*; OUP Oxford: Oxford, UK, 2012.

22. Ellis, R. *A Commentary on Catullus*; Adamant Media Corporation: Tampa, FL, USA, 2005.
23. Kelly, J.F. *The Ecumenical Councils of the Catholic Church: A History*; Liturgical Press: Collegeville, MN, USA, 2009; 232p.
24. Pope Gregory XIII. Inter Gravissimas. 1582. Available online: https://myweb.ecu.edu/mccartyr/inter-grav.html (accessed on 20 November 2023).
25. Cappelli, A. *Cronologia, Cronografia e Calendario Perpetuo, a Cura di Marino Vigano*, 7th ed.; HOEPLI EDITORE: Milano, Italy, 1998.
26. Sigismondi, C. Algoritmi per il calcolo dell'epatta della Luna. *Gerbertvs* **2016**, *9*, 109–112.
27. NASA's SVS. Moon Phase and Libration, 2020; Home-NASA Scientific Visualization Studio. 2021. Available online: https://svs.gsfc.nasa.gov/4768/ (accessed on 20 November 2023).
28. Bergeron, J. *Highlights of Astronomy: As Presented at the XXIst General Assembly of the IAU, 1991*; Springer Science & Business Media: Berlin, Germany, 2013.
29. Ratkowski, R.; Foster, J. Libration of the Moon. In *Earth Science Picture of the Day*; 2014. Available online: https://epod.usra.edu/blog/2014/05/libration-of-the-moon.html (accessed on 30 January 2024).
30. Losee, J. *Filosofia Della Scienza*; Un'introduzione; Il Saggiatore: Milano, Italy, 2009.
31. Drake, S. La mela di Newton ed il Dialogo di Galileo. *Le Scienze* **1980**, *146*, 90.
32. Koyré, A. *Studi Galileiani*; G. Einaudi: Torino, Italy, 1979; 355p.
33. Galilei, G. Sidereus Nuncius, a cura di A. Battistini, Venezia, Marsilio. *Nuncius* **1993**, *8*, 720–722.
34. Barbera, G. *Le Opere, Edizione Nazionale*; Il Saggiatore: Firenze, Italy, 1896; Volume VI.
35. Piccolino, M.; Wade, N.J. *Galileo's Visions*; Oxford University Press: Oxford, UK, 2013.
36. Danjon, A. Relation Entre l'Eclairement de la Lune Eclipsee et l'Activite Solaire. *L'Astronomie* **1921**, *35*, 261–265.
37. Fotografie Lunari e Degli Altri Corpi Celesti. In *Memorie dell'Osservatorio del Collegio Romano D.C.D.G.*; Nuova serie dell'anno 1857 al 1859; Tipografia delle Belle Arti: Roma, Italy, 1859.
38. *Sui Recenti Progressi Della Meteorologia*; Tipografia delle Belle Arti: Roma, Italy, 1861.
39. *Unità Delle Forze Fisiche*; Tipografia Forense: Roma, Italy, 1864.
40. *Le Soleil: Exposé des Principales Découvertes Modernes*; Librairie Alain Brieux: Paris, France, 1870.
41. Secchi, A. *Rapporto della Commissione per la Misura del Meridiano Centrale Europeo Negli Stati Pontifici*; Tip. delle scienze matematiche e fisiche: Roma, Italy, 1871.
42. Skylab. Available online: https://www.nasa.gov/skylab/ (accessed on 30 January 2024).
43. SOHO. Available online: https://science.nasa.gov/mission/soho (accessed on 30 January 2024).
44. Available online: https://soho.nascom.nasa.gov/ (accessed on 30 January 2024).
45. Kimball, D.S. *A Study of the Aurora of 1859*; Geophysical Institute at the University of Alaska: Fairbanks, AK, USA, 1960.
46. Carrington, R.C. Description of a singular appearance seen in the Sun on September 1, 1859. *Mon. Not. R. Astron. Soc.* **1859**, *20*, 13–15. [CrossRef]
47. Hodgson, R. On a curious appearance seen in the Sun. *Mon. Not. R. Astron. Soc.* **1859**, *20*, 15–16. [CrossRef]
48. Maynard, T.; Smith, N.; Gonzalez, S. *Solar Storm Risk to the North American Electric Grid*; Lloyd's of London and Atmospheric and Environmental Research, Inc.: London, UK, 2013.
49. Ermolli, I.; Ferrucci, M. Registro di disegni del disco solare, "31 agosto 1859". In *Tra cielo e Terra: L'avventura scientifica di Angelo Secchi*; Comitato Nazionale per il Bicentenario della Nascita di Angelo Secchi: Rome, Italy, 2018.
50. Baker, D.; Balstad, R.; Bodeau, M.; Cameron, E.; Fennell, J.; Fisher, G.; Forbes, K.; Kintner, P.; Leffler, L.; Lewis, W.; et al. *Severe Space Weather Events – Understanding Societal and Economic Impacts*; The National Academy Press: Washington, DC, USA, 2008.
51. Xu, Z.-T. The hexagram "Feng" in "the book of changes" as the earliest written record of sunspot. *Chin. Astron.* **1980**, *4*, 406.
52. Vaquero, J.M. Letter to the Editor: Sunspot observations by Theophrastus revisited. *J. Br. Astron. Assoc.* **2007**, *117*, 346.
53. Carlowicz, M.J.; Lopez, R. *Storms from the Sun: The Emerging Science of Space Weather*; Joseph Henry Press: Washington, DC, USA, 2002.
54. Scafetta, N. Solar Oscillations and the Orbital Invariant Inequalities of the Solar System. *Sol. Phys.* **2020**, *295*, 33. [CrossRef]
55. Dunham, D.W.; Dunham, J.B. Observing Total Solar Eclipses from Near the Edge of the Predicted Path. *Moon* **1973**, *8*, 546. [CrossRef]
56. Sofia, S.; O'Keefe, J.; Lesh, J.R.; Endal, A.S. Solar Constant: Constraints on Possible Variations Derived from Solar Diameter Measurements. *Science* **1979**, *204*, 1306. [CrossRef] [PubMed]
57. Sigismondi, C.; Dunham, D.; Guhl, K.; Andersson, S.; Bode, H.; Canales, O.; Colona, P.; Farago, O.; Fernández-Ocaña, M.; Gabel, A.; et al. Baily's Beads Atlas in 2005–2008 Eclipses. *Sol. Phys.* **2009**, *258*, 191–202. [CrossRef]
58. Raponi, A.; Sigismondi, C.; Guhl, K.; Nugent, R.; Tegtmeier, A. The measurement of solar diameter and limb darkening function with the eclipse observations. *Sol. Phys.* **2012**, *278*, 269–283. [CrossRef]
59. Dicke, R.H.; Goldenberg, H.M. Solar oblateness and general relativity. *Phys. Rev. Lett.* **1967**, *18*, 313. [CrossRef]
60. Hill, H.A.; Stebbins, R.T.; Oleson, J.R. The finite Fourier transform definition of an edge on the solar disk. *Astrophys. J.* **1975**, *200*, 484–498. [CrossRef]
61. Rozelot, J.P.; Damiani, C. History of solar oblateness measurements and interpretation. *Eur. Phys. J. H* **2011**, *36*, 407–436. [CrossRef]
62. Available online: https://www.icra.it/solar/pub/eclipse%20egypt.pdf (accessed on 30 January 2024).
63. Sigismondi, C. The Quest of Solar Variability. In Proceedings of the INW25 Pescara Meeting, Pescara, Italy, 8–18 July 2008.

64. Sofia, S.; Girard, T.M.; Sofia, U.J.; Twigg, L.; Heaps, W.; Thuillier, G. Variation of the diameter of the Sun as measured by the Solar Disk Sextant (SDS). *Mon. Not. R. Astron. Soc.* **2013**, *436*, 2151–2169. [CrossRef]
65. Auwers, A. Der Sonnendurchmesser und der Venusdurchmesser nach den Beobachtungen an den Heliometern der deutschen Venus-Expeditionen. *Astron. Nach.* **1891**, *128*, 361. [CrossRef]
66. Lamy, P.; Prado, J.Y.; Floyd, O.; Rocher, P.; Faury, G.; Koutchmy, S. A Novel Technique for Measuring the Solar Radius from Eclipse Light Curves–Results for 2010, 2012, 2013, and 2015. *Sol. Phys.* **2015**, *290*, 2617–2648. [CrossRef]
67. Guhl, K. Baily's Beads Observation during the Hybrid Solar Eclipse 2023 April 20. *J. Occult. Astron.* **2023**, *4*, 12–15.
68. Guhl, K. (IOTA/ES, Berlin, Germany); Quaglia, L. (University of Bologna, Bologna, BO, Italy). Personal communication, 2023.
69. Boscardin, S. Um Ciclo de Medidas do Semidiametro Solar Com Astrolabio. PhD Thesis, Observatorio Nacional, Rio de Janeiro, Brazil, 2011.
70. Sigismondi, C.; Andrei, A.H.; Boscardin, S.; Penna, J.L.; Reis-Neto, E. Optical deformations in solar glass filters for high precision solar astrometry with astrolabes and heliometers. In Proceedings of the Fourteenth Marcel Grossmann Meeting On Recent Developments in Theoretical and Experimental General Relativity, Astrophysics, and Relativistic Field Theories: Proceedings of the MG14 Meeting on General Relativity, Rome, Italy, 12–18 July 2015; pp. 3696–3701.
71. Reis-Neto, E. Observacoes Solares: Estudo das Variacoes do Diametro e Suas Correlacoes. Ph.D Thesis, Observatorio Nacional, Rio de Janeiro, Brazil, 2010.
72. Thuillier, G.; Zhu, P.; Shapiro, A.I.; Sofia, S.; Tagirov, R.; van Ruymbeke, M.; Perrin, J.-M.; Sukhodolov, T.; Schmutz, W. Solar disc radius determined from observations made during eclipses with bolometric and photometric instruments on board the PICARD satellite. *Astron. Astrophys.* **2017**, *603*, A28. [CrossRef]
73. Sigismondi, C. Docteur en Sciences. Ph.D. Thesis, Université de Nice-Sophia Antipolis, Nice, France, 2011.
74. Sigismondi, C. *Solar Activity and its Magnetic Origin IAUS 233*; Bothmer, W., Ed.; Cambridge University Press: Cambridge, UK, 2006; p. 521.
75. Sigismondi, C. *Gerbertus*; 2022; Volume 18, 192p. Available online: https://www.icra.it/gerbertus/2022/Gerbertus18.pdf (accessed on 30 January 2024).
76. Hale, G.E. On the probable existence of a magnetic field in sun-spots. *Astrophys. J.* **1908**, *28*, 315. [CrossRef]
77. Finsterle, W.; Jefferies, S.M.; Cacciani, A.; Rapex, P.; Giebink, C.; Knox, A.; DiMartino, V. Seismology of the solar atmosphere. *Sol. Phys.* **2004**, *220*, 317–331. [CrossRef]
78. Maunder, W. Greenwich, Royal Observatory, the "great" magnetic storms, 1875 to 1903, and their association with sun-spots. *Mon. Not. R. Astron. Soc.* **1904**, *64*, 224. [CrossRef]
79. Shapiro, I.I. Is the Sun shrinking? *Science* **1980**, *208*, 51–53. [CrossRef] [PubMed]
80. Kuhn, J.R.; Bush, R.; Emilio, M.; Scholl, I.F. The precise solar shape and its variability. *Science* **2012**, *337*, 1638–1640. [CrossRef] [PubMed]
81. Meftah, M.; Corbard, T.; Irbah, A.; Morand, F.; Ikhlef, R.; Renaud, C.; Hauchecorne, A.; Assus, P.; Chauvineau, B.; Crepel, M.; et al. PICARD SOL, a new ground-based facility for long-term solar radius measurements: First results. *J. Phys. Conf. Ser.* **2013**, *440*, 012003. [CrossRef]
82. Sigismondi, C. Measure the Solar Diameter with a Smartphone's Ghost Images. In Proceedings of the Eclispe Working Group AAS Meeting, San Antonio, TX, USA, 29–30 September 2023.

MDPI

St. Alban-Anlage 66

4052 Basel

Switzerland

www.mdpi.com

Universe Editorial Office

E-mail: universe@mdpi.com

www.mdpi.com/journal/universe